Graph Theory

and

Complex Networks

An Introduction

Maarten van Steen

Copyright © 2010 Maarten van Steen
Published by Maarten van Steen
ISBN: 978-90-815406-1-2

Edition: 1. Printing: 01 (April 2010)

All rights to text and illustrations are reserved by Maarten van Steen. This work may not be copied, reproduced, or translated in whole or part without written permission of the publisher, except for brief excerpts in reviews or scholarly analysis. Use with any form of information storage and retrieval, electronic adaptation or whatever, computer software, or by similar or dissimilar methods now known or developed in the future is strictly forbidden without written permission of the publisher.

To Mariëlle, Max, and Elke

Contents

Preface		**ix**
1	**Introduction**	**1**
	1.1 Communication networks	4
	Historical perspective	4
	From telephony to the Internet	6
	The Web and Wikis	8
	1.2 Social networks	9
	Online communities	9
	Traditional social networks	10
	1.3 Networks everywhere	11
	1.4 Organization of this book	13
2	**Foundations**	**17**
	2.1 Formalities	18
	Graphs and vertex degrees	18
	Degree sequence	23
	Subgraphs and line graphs	28
	2.2 Graph representations	31
	Data structures	31
	Graph isomorphism	33
	2.3 Connectivity	37
	2.4 Drawing graphs	45
	Graph embeddings	45
	Planar graphs	50
3	**Extensions**	**55**
	3.1 Directed graphs	57
	Basics of directed graphs	57

	Connectivity for directed graphs	61
3.2	Weighted graphs	65
3.3	Colorings	69
	Edge colorings	69
	Vertex colorings	71

4 Network traversal — 79
- 4.1 Euler tours 81
  - Constructing an Euler tour 82
  - The Chinese postman problem 87
- 4.2 Hamilton cycles 92
  - Properties of Hamiltonian graphs 92
  - Finding a Hamilton cycle 97
  - Optimal Hamilton cycles 100

5 Trees — 105
- 5.1 Background 107
  - Trees in transportation networks 107
  - Trees as data structures 109
- 5.2 Fundamentals 112
- 5.3 Spanning trees 116
- 5.4 Routing in communication networks 119
  - Dijkstra's algorithm 120
  - The Bellman-Ford algorithm 123
  - A note on algorithmic performance 127

6 Network analysis — 131
- 6.1 Vertex degrees 133
  - Degree distribution 134
  - Degree correlations 136
- 6.2 Distance statistics 140
- 6.3 Clustering coefficient 143
  - Some effects of clustering 143
  - Local view 144
  - Global view 146
- 6.4 Centrality 150

7 Random networks — 155
- 7.1 Introduction 157
- 7.2 Classical random networks 158
  - Degree distribution 159
  - Other metrics for random graphs 162

	7.3	Small worlds 167
	7.4	Scale-free networks 172
		Fundamentals 173
		Properties of scale-free networks 178
		Related networks 182

8 Modern computer networks — 185

- 8.1 The Internet 187
  - Computer networks 187
  - Measuring the topology of the Internet 192
- 8.2 Peer-to-peer overlay networks 195
  - Structured overlay networks 196
  - Random overlay networks 204
- 8.3 The World Wide Web 212
  - The organization of the Web 212
  - Measuring the topology of the Web 214

9 Social networks — 223

- 9.1 Social network analysis: introduction 225
  - Examples 225
  - Historical background 227
  - Sociograms in practice: a teacher's aid 231
- 9.2 Some basic concepts 234
  - Centrality and prestige 234
  - Structural balance 240
  - Cohesive subgroups 246
  - Affiliation networks 252
- 9.3 Equivalence 254
  - Structural equivalence 255
  - Automorphic equivalence 258
  - Regular equivalence 259

Conclusions — 261

Mathematical notations — 267

Index — 271

Bibliography — 279

PREFACE

When I was appointed Director of Education for the Computer Science department at VU University, I became partly responsible for revitalizing our CS curriculum. At that point in time, mathematics was generally experienced by most students as difficult, but even more important, as being irrelevant for successfully completing your studies. Despite numerous efforts from my colleagues from the Mathematics department, this view on mathematics has never really changed. I myself obtained a masters degree in Applied Mathematics (and in particular Combinatorics) before switching to Computer Science and gradually moving into the field of large-scale distributed systems. My own research is by nature highly experimental, and being forced to handle large systems, bumping into the theory and practice of complex networks was almost inevitable. I also never quite quit enjoying material on (combinatorial) algorithms, so I decided to run another type of experiment.

The experiment that eventually led to this text was to teach graph theory to first-year students in Computer Science and Information Science. Of course, I needed to explain why graph theory is important, so I decided to place graph theory in the context of what is now called network science. The goal was to arouse curiosity in this new science of measuring the structure of the Internet, discovering what online social communities look like, obtain a deeper understanding of organizational networks, and so on. While doing so, teaching graph theory was just part of the deal.

No appropriate book existed, so I started writing lecture notes. As with most experiments that I participate in (the hard work is actually done by my students), things got a bit out of hand and I eventually found myself writing another book. Considering that my other textbooks are really on (distributed) computer systems and barely contain any mathematical symbols (as, in fact, is also the case for most of my research papers), this book is to be considered as somewhat exceptional. In fact, because I do not consider myself to be a mathematician anymore, I'm not quite sure how this book should be classified. Is it math? Is it computer science? Does it matter?

The goal is to provide a first introduction into complex networks, yet in a more or less rigorous way. After studying this material, a student should have a pretty good idea of what makes real-world networks complex instead of complicated, and can do a lot more than just handwaving when it comes to explaining real-world phenomena. While getting to that point, I also hope to have achieved two other goals: successfully teaching the foundations of graph theory, and even more important, lowering the threshold for studying mathematical material.

The latter may not be obvious when skimming through the text: it is full of mathematical symbols, theorems, and proofs. I have deliberately chosen for this approach, feeling confident that if enough and targeted attention is paid to the language of mathematics in the first chapters, a student will become aware of the fact that mathematical language is sometimes only intimidating: mathematicians' barks are often worse than their bites. Students who have so far followed my classes have indeed confirmed that they were surprised at how much easier it was to access the math once they got over the notations. I hope that this approach will last for long, making it at least easier for many students to not immediately pull back when encountering mathematical language in other texts.

Intended readership

This book has been written for first- or second-year undergraduates who have taken the usual courses in mathematics as taught in high school. However, although I claim that the material is not inherently difficult, it will certainly require serious studying by most students, and certainly those for which math does not come natural. As mentioned, I have deliberately chosen to use the language of math because it is not only precise and comprehensive, but above all because I believe that at the level of this book, it will lower the threshold for other mathematical texts. It should be clear that the lecturer using this material may need to pay some special effort to encourage students. For most students, the language will turn out to be the hard part, not the content.

Supplementary material

As said, this book is part of a course on graph theory and complex networks. Although it can be used for self-study, I encourage students and their instructors to visit the accompanying Web site:

```
http://www.distributed-systems.net/gtcn/
```

where lots of extra material can be found, including, most importantly, a huge collection of exercises (with solutions). My goal is to expand this set of exercises continuously. This is the most important reason not to have included any exercises in the book: they can be readily obtained from the site, and always up-to-date.

To make the material more accessible (and fun), but also to allow students to do some basic analysis of larger graphs and networks, we have been using Mathematica in combination with Combinatorica. All material, including Mathematica notebooks and data on graphs are all available through the Web site. The site also has some extra tools for generating graphs.

Of course, slides and handouts are available (all originating from LaTeX sources), as well as all the figures from the book. Perhaps most importantly, an electronic version of the book itself is also available.

All material is freely accessible

Sometimes when you write a book, it makes a lot of sense to think big and act commercially. Thinking big in this sense means you expect many people to have access to your book. Acting commercially means that you try to successfully market and sell your book. Sometimes, it's enough to just think big, knowing that acting commercially will certainly keep everything small. When you write a book containing mathematical symbols, thinking big and acting commercially doesn't seem the right combination. I merely hope to see the material to be used by many students and instructors everywhere and to receive a lot of constructive feedback that will lead to improvements. Acting commercially has never been one of strong points anyway.

However, freely accessible doesn't mean that everyone has the right to copy and spread the material, which I would find quite offensive. For this reason, when requesting an electronic copy, the book will be watermarked with your e-mail address. The watermark is part of the LaTeX source, so it's pretty difficult to remove, although I do not have the illusion that removal is impossible.

Finally, for those who still prefer to (also) have a hard-copy version of the book (of course, without a watermark), such can be realized by placing an order through the Web site. Further information can be found there. The price is comparable to printing it yourself.

Acknowledgments

There are a few people who deserve to be mentioned. Spyros Voulgaris has been responsible for creating homework assignments, preparing Mathematica notebooks, and setting up all the exercise classes. Albana Gaba has a gifted talent to provide very constructive feedback (next to the fact that she has been working like a dog to process all the student assignments). Achraf Belmokadem has done a terrific job on setting up a Web-based subsystem for letting students self-assess their abilities for solving graph problems. Finally, I would like to thank the students who have undergone my teaching for the past two years and who have, despite all the mistakes, continued to claim that they enjoyed it.

Maarten van Steen
Amsterdam, April 2010

CHAPTER 1

INTRODUCTION

On 11 September 2001 there was a malicious attack on the WTC towers in New York City, eventually leading to the two buildings collapsing. What is not known to many people, is that there were three transatlantic Internet cables coming ashore close to the WTC and that an important Internet switching station was damaged, along with two other important Internet resource centers. Peter Salus and John Quarterman [2002] had since long been measuring the performance of the Internet by checking the reachability of a fairly large collection of servers. In effect, they simply sent messages from different locations on the Internet to these special computers and recorded whether or not servers would be responding. If reachability was 100%, this meant that all servers were up and running. If reachability was less, this could mean that servers were either out-of-order, or that the communication paths to some of the servers were broken.

Immediately after the attack reachability dropped by about 9%. Within 30 minutes it had almost reached its old value again.

This example illustrates two important properties of the Internet. First, even when disrupting what would seem as a vital location in the Internet, such a disruption barely affects the overall communication capabilities of the network. Second, the Internet has apparently been designed in such a way that it takes almost no time to recover from a big disaster. This recovery is even more remarkable when you consider that no manual repairs had even started, but also that no designer had ever really anticipated such attacks (although robustness was definitely a design criterion for the Internet). The Internet demonstrated emergent self-healing behavior.[1]

The Internet is an example of what is now commonly referred to as a **complex network**, which we can informally define as large collection of interconnected nodes. A node can be anything: a person, an organization, a computer, a biological cell, and so forth. Interconnected means that two nodes may be linked, for example, because two people know each other, two organizations exchange goods, two computers have a cable connecting the two of them, or because two neurons are connected by means of a synapses for passing signals. What makes these networks complex is that they are generally so huge that it is impossible to understand or predict their overall behavior by looking into the behavior of individual nodes or links.

As it turns out, complex networks are everywhere. Or, to be more precise, it turns out that if we model real-world situations in terms of networks, we often discover new things. What is striking, is that many real-world networks look alike: the structure of the Internet resembles the organization of our brain, but also the organization of online social communities. Where

---

[1] As we'll encounter in later chapters, there's no magic here: so-called routing algorithms simply adjust their decisions when paths break.

these similarities come from is still a mystery, just as it is often very difficult to understand how certain networks were actually structured. Before we go deeper into what complex networks actually entails, let's first consider a few general areas where networks play a vital role, starting with communication networks.

1.1 Communication networks

Not even so long ago, setting up a phone call to someone on the other side of the world required the intervention of a human operator. Moreover, an established connection was no guarantee for being able to understand each other as the quality could be pretty bad. Many will recall these situations to happen in the 70s and 80s of the previous century—really not that long ago. Today, cell phones allow us to be contacted virtually anywhere and anytime, and coverage continues to expand to even the most remote areas. Setting up a high-quality voice connection over the Internet with peers anywhere around the world is plain simple. Along these lines, we need merely wait a while until it is also possible to have cheap, high-quality video connections allowing us to experience our remote friends as being virtually in the same room.

The world appears to be becoming smaller, and people are becoming ever more connected. Obviously, telecommunication has played a crucial role in establishing this **connected world** as it is commonly known, but with the convergence of telecommunication and data networks (and notably the Internet), it is difficult not to be connected anymore. Being connected has profound effects for the dissemination of information. And as we shall see, *how* we are connected plays a crucial role when it comes to the speed and robustness of such dissemination, among many other issues.

Historical perspective

To have a connected world it is obvious that we need to communicate. If we want this world to have significant coverage, long-distance communication is obviously important. Unlike what many tend to believe, networks that facilitate such communication have a long history, as described by Holzmann and Pehrson [1995]. Apart from well-known means of communication such as sending messengers or using pigeons, long-distance communication without the need to physically transport a message has always caught the attention of mankind. Typically, such **telegraphic communication** used to be done through fire beacons, mirrors (i.e., *heliographic communication*), drums, and flags. Communication paths set up using such methods, for example by having communication posts organized at line-of-sight distances, are known from Greek and Roman history.

However, it wasn't until the end of the 18th Century that a systematic approach was developed to establish telegraphic communication *networks*. Such networks would consist of communication posts, of which pairs would lie in each other's line-of-sight. Typically, for these optical telegraphs, distances between two posts would be in the order of tens of kilometers, which was realistic given that high-quality telescopes could be used. An important aspect in the design of these networks was the **communication protocol**, which would prescribe the encoding of letters, but also what to do if there was a transmission error. To make matters more concrete, consider Figure 1.1 which shows a model of a **shutter telegraph**.

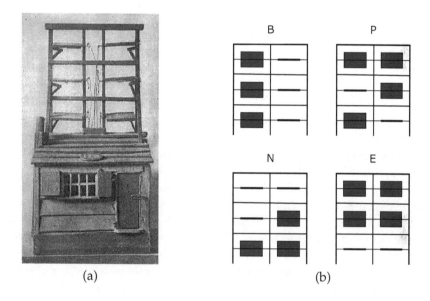

Figure 1.1: (a) A model of a shutter station with six (open) shutters and (b) a few examples of how letters were encoded.

As shown in Figure 1.1(b), letters are represented by specific combinations of open and closed shutters. In this way, it became possible to transmit messages over long distances. Of course, it became equally important to think about encryption of messages, handling transmission errors, synchronization between transmitter and reader (i.e., sender and receiver), and so on. In other words, these seemingly primitive communication networks had to deal with virtually the same issues as modern systems. Conceptually, there is really no difference.

By the middle of the 19th Century, Europe had optical telegraphic networks installed in the Scandinavian countries, France, England, Germany, and others. Concerning topology, these networks were relatively simple: there were only relatively few nodes (i.e., communication posts), and cycles did not exist. That is, between any two nodes messages could travel only through a unique path. Such networks are also known as **trees**.

Matters became serious when the electrical telegraph system emerged. Instead of using vision, communication paths were realized through electrical cables. The medium proved to be successful: by the middle of the 19th Century the electrical telegraph spanned more than 30,000 kilometers in the United States, making it more than just a serious competitor to optical telegraph systems. In fact, by then it was clear to most people that the optical networks were heading towards a dead end. In 1866, networks in the United States and Europe were successfully connected through a transatlantic cable (where earlier attempts had failed). Gradually, the concept of a worldwide network was becoming reality.

From telephony to the Internet

The impact of a worldwide telephony network can only be underestimated. From an end user's perspective, it really didn't matter anymore where you were, but only that the other party was simultaneously online. In other words, telecommunication networks realized *location independency*. This independency could be realized only because it was possible to establish a circuit between the two communicating parties: a **communication path** from one party to the other with intermediate nodes operating as switches. In most cases, these switches had fixed locations and every switch was physically linked to a few other switches. The combination of switches and links form a **communication network**, which can be represented mathematically by what is known as a **graph**, the object of study in this book.

As we already discussed, telecommunication networks were well established when people began to think about connecting computers and thus establishing data communication networks. Of course, the many existing networks already made it possible to send data, for example, as a telegram. The new challenge was to connecting these separate networks into logically a single one that could be used by computers using the same **protocol**. This led to the idea of building a communication system in which possibly large messages were split into smaller units called packets. Each packet would be tagged with the address of its destination and subsequently **routed** through the various networks. It is important to note that packets from the same message could each follow their own route to the destination, where they would then be subsequently used to reassemble the original message.

When a switch received a packet, it would only then decide to which next switch the packet would be forwarded. This **packet switching** approach contrasts sharply with telecommunication networks in which two end points would first establish a path and then subsequently let all communication pass through that path, also referred to as **circuit switching**.

The first packet-switching network was established in 1969, called the ARPANET (Advanced Research Projects Agency Network). It formed the starting point of the present Internet. Key to this network were the **interface message processors** (IMPs), special computers that provided a system-independent interface for communication. In this way, any computer that wanted to hook up to the ARPANET needed only to conform to the interface of an IMP. IMPs would then further handle the transfer of packets. They formed the first generation of network switches, or **routers**. To give an impression of what this network looked like, Figure 1.2 shows a logical map of IMPs and their connected computers as of April 1971.

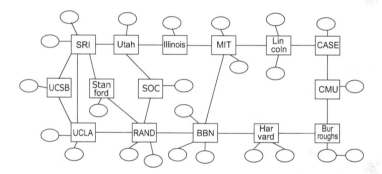

Figure 1.2: A map of the ARPANET as of April 1971. Rectangles represent IMPs; ovals are computers.

The ARPANET of 1971 constituted a network with 15 nodes and 19 links. It is so small that we can easily draw it. We've passed that stage for the Internet. (In fact, it is far from trivial to determine the size of today's Internet.) Of course, that network was also **connected**: it is possible to route a packet from any source to any destination. In fact, connectivity could still be established if a randomly selected single link broke. An important design criterion for communication networks is how many links need to fail before the network is partitioned into several parts. For our example network of Figure 1.2, it is clear that this number is 2. Rest assured that for the present-day Internet, this number is *much* higher.

Likewise, we can ask ourselves how many nodes (i.e., switches or IMPs) need to fail before connectivity is affected. Again, it can be seen that we need to remove at least 2 nodes before the network is partitioned. Surprisingly, in the present-day Internet we need not remove that many nodes to establish the same effect. This is caused by the structure of the Internet: researchers have discovered that there are relatively few nodes with very many links. These nodes essentially form an Achilles' heel of the Internet. In subsequent chapters, you will learn why.

The Web and Wikis

Next to the importance of e-mail and other Internet messaging systems, there is little discussion about the impact of the World Wide Web. The Web is an example of a digital **information space**: a collection of units of information, linked together into a network. The Web is perhaps the biggest information space that we know of today: by the end of January 2005, it was estimated to have at least 11.5 billion indexable pages [Gulli and Signorini, 2005], that is, pages that could be found and indexed by the major search engines such as Google. Three years later, different studies (using different metrics) indicate that we may be dealing with 30-50 billion pages. In any case, we are clearly dealing with a phenomenal growth.

What makes information spaces such as the Web interesting for our studies, is that again these spaces form a network. In the case of the Web, each page may (and generally will) contain links to other pages and corresponds to a node in the network. What becomes interesting are questions such as:

- If we take the number of links pointing to a page as a measure of that page's popularity, what can we say about the number and intensity of page popularity (i.e., what is the *distribution* of page popularity)?

- Does the Web also share characteristics with what are known as **small world networks**: is it possible to navigate to any other page through only a few links?

As we shall discuss extensively in Chapter 8, the Web indeed has its own characteristics, some of which correspond to those in small worlds. However, there are also important differences. For example, it turns out that the distribution of page popularity is very skewed: there are relatively few, but extremely popular pages. In contrast, by far most pages are not popular, yet there are *many* of such unpopular pages, which makes the collection of unpopular pages by itself and interesting subject for study.

An information space related to the Web is that of the online encyclopedia Wikipedia. By the end of 2007, over 7.5 million pages were counted, written in more than 250 different languages. The English Wikipedia is by far the largest, with more than 2 million articles. It is also the most popular one when measuring the number of page requests: 45% of all Wikipedia traffic is directed towards the English version [Urdaneta et al., 2009]. Again, Wikipedia forms a network with its pages as nodes and references to other pages as links. Like the Web, it turns out that there are few very popular pages, and many unpopular ones (but so many that they cannot be ignored) [Voss, 2005].

1.2 Social networks

Next to communication networks, networks that are built around people have since long been subject of study. We first consider modern social networks that have come into play as online communities facilitated by the Internet.

Online communities

In their landmark essay, Licklider and Taylor [1968] foresaw that computers would form a major communication device between people leading to the online communities much like the ones we know today. Indeed, perhaps one of the biggest successes of the Internet has been the ability to allow people to exchange information with each other by means of user-to-user messaging systems [Wams and van Steen, 2004]. The best known of these systems is e-mail, which has been around ever since the Internet came to life. Another well-known example is network news, through which users can post messages at electronic bulletin boards, and to which others may subsequently react, leading to discussion threads of all sorts and lengths. More recently instant messaging systems have become popular, allowing users to directly and interactively exchange messages with each other, possibly enhanced with information on various states of presence.

It is interesting to observe that from a technological point of view, most of these systems are really not that sophisticated and are still built with technology that has been around for decades. In many ways, these systems are simple, and have stayed simple, which allowed them to scale to sizes that are difficult to imagine. For example, it has been estimated that in 2006 almost 2 million e-mail messages were sent *every second*, by a total of more than 1 billion users. Admittedly, more than 70% of these messages were spam or contained viruses, but even then it is obvious that a lot of online communication took place. These numbers continue to rise.

More than the technology, it is interesting to see what these communication facilities do to the people who use them. What we are witnessing today is the rise of **online communities** in which people who have never met each other physically are sharing ideas, opinions, feelings, and so on. In fact, Dodds et al. [2003] have shown that also for online communities we are dealing with what is known as a **small world**. To put it simply, a small world is characterized by the fact that every two people can reach each other through a chain of just a handful of messages. This phenomenon is also known as the "six degrees of separation" [Watts, 2003] to which we will return extensively later.

Dodds et al. were interested to see whether e-mail users were capable of sending a message to a specific person without knowing that person's address. In that case, the only thing you can do is send the message to one of your acquaintances, hoping that he or she is "closer" to the target than you are. With over 60,000 users participating in the experiment, they found that 384 out of the approximately 24,000 message chains made it to designated target people (there were 18 targets from 13 different countries all over the world). Of these 384 chains, 50% had a length smaller than 5–7, depending on whether the target was located in the same country as where the chain started.

What we have just described is the phenomenon of messages traveling through a network of e-mail users. Users are linked by virtue of knowing each other, and the resulting network exhibits properties of small worlds, effectively connecting every person to the others through relatively small chains of such links. Describing and characterizing these and other networks forms the essence of **network science**.

Traditional social networks

Long before the Internet started to play a role in many people's lives, sociologists and other researchers from the humanities have been looking at the structure of groups of people. In most cases, relatively small groups were considered, necessarily because analysis of large groups was often not feasible.

An important contribution to social network analysis came from Jacob Moreno who introduced **sociograms** in the 1930s. A sociogram can be seen as a graphical representation of a network: people are represented by dots (called **vertices**) and their relationships by lines connecting those dots (called **edges**). An example we will come across in Chapter 9 is one in which a class of children are asked who they like and dislike. It is not hard to imagine that we can use a graphical representation to represent who likes whom, as shown in Figure 1.3.

Decades later, under the influence of mathematicians, sociograms and such were formalized into **graphs**, our central object of study. As mentioned, graphs are mathematical objects, and as such they come along with

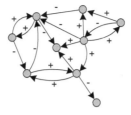

Figure 1.3: The representation of a sociogram expressing affection between people. The absence of a link indicates neutrality.

a theoretical framework that allows researchers to focus on the *structure* of networks in order to make statements about the behavior of an entire social group.

Social network analysis has been important for the further development of graph theory, for example with respect to introducing metrics for identifying importance of people or groups. For example, a person having many connections to other people may be considered relatively important. Likewise, a person at the center of a network would seem to be more influential than someone at the edge. What graph theory provides us are the tools to formally describe what we mean by relatively important, or having more influence. Moreover, using graph theory we can easily come up with alternatives for describing importance and such. Having such tools has also facilitated being more precise in statements regarding the position or role that person has within a community. We will come across such formalities in Chapter 9.

1.3 Networks everywhere

Communication networks and social networks are two classes of networks that many people are aware of. However, there are many more networks as shown in Figure 1.4. What should immediately become clear is that networks occur in very different scientific disciplines: economics, organizational studies, social sciences, biology, logistics, and so forth. What's more, the terminology that is used to describe the different networks in each discipline is largely the same, which makes it relatively easy for members of different communities to cooperate in understanding the foundations of complex networks. What is even more striking is the fact that networks from very different disciplines often look so much alike. This common terminology and the strong resemblance of networks across scientific disciplines has been instrumental in boosting network science.

Network	Vertices	Edges	Description
Airline transportation	airports	flights	Consider the scheduled flights (of a specific) carrier between two airports.
Street plans	junctions	road segment	A road segment extends exactly between two junctions. A variation is to distinguish between one-way and two-way segments.
Train transportation	stations	connection	Two stations are connected only if there is a train connection scheduled that does not pass (possibly without stopping) any intermediate stations.
Railway network	junctions	track segment	Consider the actual railway tracks. Where track segments merge or cross, we have junctions.
Brain	neurons	synapses	Each neuron can be considered to consist of inputs (called dendrites) and outputs (called axon). Synapses carry electrical signals between neurons.
Genetic networks	genes	transcription factor	In genetic (regulatory) networks we model how genes influence each other, in particular, how the product of one gene determines the rate at which another gene is transcribed (i.e., at which rate it produces its own output).
Ant colonies	junctions	pheromone trails	In order for ants to tell each other where sources of food are, they produce pheromones which is a chemical that can be picked up by other ants. Pheromones jointly constitute paths.
Citation networks	authors	citation	In scientific literature, it is common practice to (extensively) refer to related published work and sources of statements, in turn leading to citation networks.
Telephone calls	number	call	Networks of phone calls reflect (mostly) pairs of people exchanging information, thus forming a social network technically represented by phone numbers and actual calls.
Reputation networks	people	rating	In electronic trading networks such as e-Bay, buyers rate transactions. As buyers in turn can also be sellers, we obtain a network in which rates reflect the reputation between people.

Figure 1.4: Examples of networks.

Understanding complex networks requires the right set of tools. In our case, the tools we need come from a field of mathematics known as **graph theory**. In this book, you'll learn about the essential elements of graph theory in order to obtain insight into modern networks. Next to that, we discuss a number of concepts that are normally not found in traditional textbooks on graph theory, such as random networks and various metrics for characterizing graphs.

1.4 Organization of this book

In the following chapters we'll go through the foundations of graph theory and move on into parts that are normally discussed in more advanced textbooks on networks. The goal of this text is to provide only an awareness and basic understanding of complex networks, for which reason none of the advanced mathematics that accompany complex networks is discussed. To make matters easier, special notes are included that generally provide further information, such as the following:

> **Note 1.1** (More information)
> This is an example of how additional side notes are presented. Text in such notes can always be skipped as notes do not affect the flow of the main text.

There are different types of notes:

Study tips: Studying graph theory is not always easy, not because the material is so difficult, but because identifying the best approach to tackle a specific problem may not be obvious. I have compiled various tips based on experience in teaching (and once myself learning) graph theory. Students are strongly encouraged to read these tips and put them to their own advantage.

Mathematical language: For many people, mathematics is and remains a barrier to accessing otherwise interesting material. The language of mathematicians as well as the commonly used tools and techniques are sometimes even intimidating. However, there are so many cases in which the barrier is only virtual. The only thing that is needed is getting acquainted with some basics and learning how to apply them. In notes focusing on mathematical language, I generally take a step back on previously presented material and translate the math into plain English, explain mathematical notations, and so forth. These notes are meant to help understand the math, but do not serve as a replacement. Mathematics simply offers a level of precision that is difficult to match with (informal) English, yet the notations should not be something to keep anyone away from reaching a deeper understanding.

Proof techniques: Notably in Chapters 2 and 3 some time is taken to explain a bit more about how to prove theorems. One of the main difficulties that I experienced when first studying graph theory and more generally, combinatorics, was finding structure in proofs. As in virtually any other field of mathematics, graph theory uses a whole array of proof techniques. In these notes, the most commonly used ones are made explicit, aiming at creating a better awareness of available techniques so that students may have less of a feeling of walking in the dark when it comes to solving mathematical problems.

Algorithmics: Graph theory involves many algorithms, such as, for example, finding shortest paths, identifying reachable vertices, determining similarity, and so on. Traditionally, algorithms have always been described using math, but that language is not particularly well-equipped for expressing the flow of control inherent to most algorithms. In algorithmic notes some of graph algorithms are expressed in pseudo code, roughly following a traditional programming language. In virtually all cases, this description leads to a better separation of the actual math and the steps comprising an algorithm.

More information: These type of notes contain a wide variety of information, ranging from additional background material to more difficult mathematical material such as proofs. In all cases, these notes do not interfere with the main text and may be skipped on first reading.

Proofs that have been marked "(*)" may be skipped at first reading: they are to be considered the tougher parts of the material.

The book is roughly organized into two parts. The first parts covers Chapters 2–6. These chapters roughly cover the same material that can usually be found in standard textbooks on graph theory. Except for Chapter 6, this material is to be considered essential for studying graph theory and should in any case be covered. Chapter 6 can be considered as a compilation of various metrics from different disciplines to characterize graphs, their structures, and the positions that different nodes have in networks.

The second part consists of Chapters 7–9 and discusses (graph models of) real-world networks. Notably Chapter 7 on random networks contains material that is often presented only in more advanced textbooks yet which I consider to be crucial for raising scientific interest in modern network science. Random networks are important from a conceptual modeling point of view, from an analysis point of view, and are important for explaining the emergent behavior we see in real-world systems. By keeping explanations as simple as possible and attempting to stick only to the core elements, this material should be relatively easy to access for anyone having essentially learned only high-school mathematics. The two succeeding chapters discuss theory and practice of real-world systems: computer networks and social networks, respectively.

Chapter 2

Foundations

In the previous chapter we have informally introduced the notion of a network and have given several examples. In order to study networks, we need to use a terminology that allows us to be precise. For example, when we speak about the distance between two nodes in a network, what do we really mean? Likewise, is it possible to specify how well connected a network is? These and other statements can be formulated accurately by adopting terminology from **graph theory**. Graph theory is a field in mathematics that gained popularity in the 19th and 20th century, mainly because it allowed to describe phenomena from very different fields: communication infrastructures, drawing and coloring maps, scheduling tasks, and social structures, just to name a few.

We will first concentrate only on the foundations of graph theory. To this end, we will use the language of mathematics, as it allows us to be precise and concise. However, to many this language with its many symbols and often peculiar notations can easily form an obstacle to grasp the essence for what it is being used. For this reason, we will gently and gradually introduce notations while providing more verbose descriptions alongside the more formal definitions. You are encouraged to pay explicit attention to the formalities: in the end, they will prove to be much more convenient to use than verbose verbal descriptions. The latter often simply fail to be precise enough to completely understand what is going on. It is also not that difficult, as most notations come directly from set theory.

2.1 Formalities

Let us start with discussing what is actually meant by a network. To this end, we first concentrate on some basic formal concepts and notations from graph theory, together with a few fundamental properties that characterize networks. After having studied this section, you will have already learned a lot about the world of graphs and should also feel more comfortable with mathematical notations.

Graphs and vertex degrees

As said, the networks that have been introduced so far are mathematically known as graphs. In its simplest form, a graph is a collection of vertices that can be connected to each other by means of edges. In particular, each edge of graph joins exactly two vertices. Using a formal notation, a graph is defined as follows.

Definition 2.1: *A **graph** G consists of a collection V of **vertices** and a collection **edges** E, for which we write $G = (V, E)$. Each edge $e \in E$ is said to **join** two* vertices, which are called its **end points**. If e joins $u, v \in V$, we write $e = \langle u, v \rangle$. Vertex u and v in this case are said to be **adjacent**. Edge e is said to be **incident** with vertices u and v, respectively.

We will often write $V(G)$ and $E(G)$ to denote the set of vertices and edges associated with graph G, respectively. It is important to realize that an edge can actually be represented as an *unordered* tuple of two vertices, that is, its end points. For this reason, we make no distinction between $\langle v, u \rangle$ and $\langle u, v \rangle$: they both represent the fact that vertex u and v are adjacent.

This definition may already raise a few questions. First of all, is it possible that an edge joins the same vertices, that is, can an edge form a **loop**? There is nothing in the definition that prevents this, and indeed, such edges are allowed. Likewise, you may be wondering whether two vertices u and v may be joined by **multiple edges**, that is, a set of edges each having u and v as their end points. Indeed, this is also possible, and we shall be discussing a few examples shortly. A graph that does not have loops or multiple edges is called **simple**.

Likewise, there is nothing that prohibits a graph from having no vertices at all. Of course, in that case there will also be no edges. Such a trivial graph is called **empty**. Another special case is formed by a simple graph having n vertices, with each vertex being adjacent to every other vertex. This graph is also known as a **complete graph**. A complete graph with n vertices is commonly denoted as K_n. Finally, the **complement of a graph** G, denoted as $\overline{G}$ is the graph obtained from G by removing all its edges and joining exactly those vertices that were *not* adjacent in G. It should be clear that if we take a graph G and its complement $\overline{G}$ "together," we obtain a complete graph. Taking two graphs "together" will be made more precise later in this chapter.

As an aside, notice that when we write $\langle u, v \rangle$, we can say only that u and v are adajacent, that is, that there is *at least one* edge that joins the two. Strictly speaking, it is not possible using this notation to distinguish different edges that all happen to join both u and v. If we wanted to make that distinction, we would have to write something like $e_1 = \langle u, v \rangle$ and $e_2 = \langle u, v \rangle$. In other words, we would have to explicitly enumerate the edges that join u and v. Of course, when dealing with *simple* graphs, there can be no mistake about which edge we are considering when we write $\langle u, v \rangle$. Here we see an example where mathematics allows us to be precise and unambiguous. We will encounter many more of such examples.

As in so many practical situations, it is often convenient to talk about your neighbors. In graph-theoretical terms, the neighbors of a vertex u are formed by the vertices that are adjacent to v, or, in other words, those vertices to which v has been joined by means of an edge. We can formulate this precisely using formal mathematical notations as follows.

Definition 2.2: *For any graph G and vertex $v \in V(G)$, the **neighbor set** $N(v)$ of v is the set of vertices (other than v) adjacent to v, that is*
$$N(v) \stackrel{\text{def}}{=} \{w \in V(G) \mid v \neq w, \exists e \in E(G) : e = \langle u, v \rangle\}$$

Note 2.1 (Mathematical language)
The formal notation is Definition 2.2 is very precise, yet can be somewhat intimidating. Let us decypher it a bit. First, we use the symbol $\stackrel{\text{def}}{=}$ to express that what is written on the left-hand side is defined by what is written on the right-hand side. In other words,
$$N(v) \stackrel{\text{def}}{=} \ldots$$
is nothing but accurately stating that $N(v)$ is defined by what follows on the right hand of $\stackrel{\text{def}}{=}$. Recall that the symbol '$\exists$' is the **existential quantifier** used in set theory to express statements like "there exists an ..." Keeping this in mind, you should now be able to see that the right-hand side translates into English to the following statement:

> The set of vertices w in G, with w not equal to v, such that there exists an edge e in G that joins v and w.

We will be encountering many more of these formal statements. If you have trouble correctly interpreting them, we encourage you to make translations like the previous one to actually practice *reading* mathematics. After a while, you will notice that these translations come naturally by themselves.

The word "graph" comes from the fact that it is often very convenient to use a graphical representation, as shown in Figure 2.1. In this example, we have a graph G with eight vertices and a total of 18 edges. Each vertex is represented as a black dot whereas edges are drawn as lines. When drawing a graph, it is often convenient to add labels. Both vertices and edges can be labeled. We shall generally not use subscripts when labeling vertices and edges in our drawings of graphs. This means that a label such as e13 from Figure 2.1 is the same as e_{13} in our text.

It should be clear that there may be many different ways to draw a graph. In the first place, there is no reason why we would stick to just dots and lines, although it is common practice to do so. Secondly, there are, in principle, no rules concerning on where to position the drawn vertices, nor are there any rules stating that a line should be drawn in a straight fashion. However, the way that we draw graphs is often important when it comes to visualizing certain aspects. We return to this issue extensively in Section 2.4.

2.1. FORMALITIES

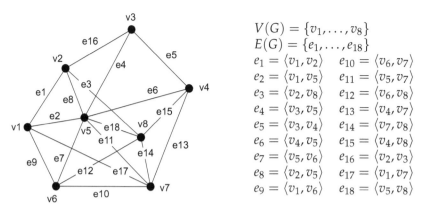

Figure 2.1: An example of a graph with eight vertices and 18 edges.

An important property of a vertex is the number of edges that are incident with it. This number is called the degree of a vertex.

Definition 2.3: *The number of edges incident with a vertex v is called the **degree** of v, denoted as $\delta(v)$. Loops are counted twice.*

Let us consider our example from Figure 2.1 again. In this case, because there are four edges incident with vertex v_1, we have that $\delta(v_1) = 4$. We can complete the picture by considering every vertex, which gives us:

Vertex	Degree	Incident edges	Neighbors
v_1	4	$\langle v_1,v_2\rangle, \langle v_1,v_5\rangle, \langle v_1,v_6\rangle, \langle v_1,v_7\rangle$	v_2, v_5, v_6, v_7
v_2	4	$\langle v_1,v_2\rangle, \langle v_2,v_3\rangle, \langle v_2,v_5\rangle, \langle v_2,v_8\rangle$	v_1, v_3, v_5, v_8
v_3	3	$\langle v_2,v_3\rangle, \langle v_3,v_4\rangle, \langle v_3,v_5\rangle$	v_2, v_4, v_5
v_4	4	$\langle v_3,v_4\rangle, \langle v_4,v_5\rangle, \langle v_4,v_7\rangle, \langle v_4,v_8\rangle$	v_3, v_5, v_7, v_8
v_5	7	$\langle v_1,v_5\rangle, \langle v_2,v_5\rangle, \langle v_3,v_5\rangle, \langle v_4,v_5\rangle, \langle v_5,v_6\rangle, \langle v_5,v_7\rangle, \langle v_5,v_8\rangle$	$v_1, v_2, v_3, v_4, v_6, v_7, v_8$
v_6	4	$\langle v_1,v_6\rangle, \langle v_5,v_6\rangle, \langle v_6,v_7\rangle, \langle v_6,v_8\rangle$	v_1, v_5, v_7, v_8
v_7	5	$\langle v_1,v_7\rangle, \langle v_4,v_7\rangle, \langle v_5,v_7\rangle, \langle v_6,v_7\rangle, \langle v_7,v_8\rangle$	v_1, v_4, v_5, v_6, v_8
v_8	5	$\langle v_2,v_8\rangle, \langle v_4,v_8\rangle, \langle v_5,v_8\rangle, \langle v_6,v_8\rangle, \langle v_7,v_8\rangle$	v_2, v_4, v_5, v_6, v_7

When adding the degrees of all vertices from G, we find that the total sum is 36, which is exactly twice the number of edges. This brings us to our first theorem:

Theorem 2.1: *For all graphs G, the sum of the vertex degrees is twice the number of edges, that is,*

$$\sum_{v \in V(G)} \delta(v) = 2 \cdot |E(G)|$$

Proof. When we count the edges of a graph G by enumerating for each vertex v of G the edges incident with that vertex v, we are counting each edge exactly twice. Hence, $\sum_{v \in G} \delta(v) = 2 \cdot |E(G)|$. □

Note 2.2 (Mathematical language)
Again, we encounter some formal mathematical notations. In this case, we use the standard symbol $\sum$ as an abbreviation for summation. Thus, $\sum_{i=1}^{n} x_i$ is the same as $x_1 + x_2 + x_3 + \cdots + x_n$. In many cases, the summation is simply over all elements in a specific set, such as in our example where we consider all the vertices in a graph. In that case, if we assume that $V(G)$ consists of the vertices $v_1, v_2, \ldots, v_n$, the notation $\sum_{v \in V(G)} \delta(v)$ is to be interpreted as:

$$\sum_{v \in V(G)} \delta(v) \stackrel{\text{def}}{=} \delta(v_1) + \delta(v_2) + \cdots + \delta(v_n)$$

Note, furthermore, that we use the notation $|S|$ to denote the size of a set S. In our example, $|E(G)|$ thus denotes the size of $E(G)$ or, in other words, the total number of edges in graph G.

There is also an interesting corollary that follows from this property, namely that the number of vertices with an odd degree must be even. This can be easily seen if we split the vertices V of a graph into two groups: V_{odd} containing all vertices with odd degree, and V_{even} with all vertices having even degree. Clearly, if we take the sum of all the degrees from vertices in V_{odd}, and those from V_{even}, we will have summed up all vertex degrees, that is,

$$\sum_{v \in V_{odd}} \delta(v) + \sum_{v \in V_{even}} \delta(v) = \sum_{v \in V} \delta(v)$$

which is even. Because the sum of even vertex degrees is obviously even, we know that $\sum_{v \in V_{even}} \delta(v)$ is even. This can only mean that $\sum_{v \in V_{odd}} \delta(v)$ must also be even. Combining this with the fact that all vertex degrees in V_{odd} are odd, we conclude that the number of vertices with odd degree must be even, that is, $|V_{odd}|$ is even. We have thus just proven:

Corollary 2.1: *For any graph, the number of vertices with odd degree is even.*

The vertex degree is a simple, yet powerful concept. As we shall see throughout this text, vertex degrees are used in many different ways. For example, when considering social networks, we can use vertex degrees to express the importance of a person within a social group. Also, when we discuss the structure of real-world communication networks such as the Internet, it will turn out that we can a learn a lot by considering the *distribution* of vertex degrees. More specifically, by simply ordering vertices by their

2.1. FORMALITIES

vertex degree, we will be able to obtain insight in how such a network is actually organized.

Degree sequence

Listing the vertex degrees of a graph gives us a **degree sequence**. The vertex degrees are usually listed in descending order, in which case we refer to an **ordered degree sequence**. For example, if we consider the eight vertices of graph G from Figure 2.1, we have the following vertex degrees

vertex:	v_1	v_2	v_3	v_4	v_5	v_6	v_7	v_8
degree:	4	4	3	4	7	4	5	5

which, when ordering these degrees in descending order, leads to the ordered degree sequence

$$[7, 5, 5, 4, 4, 4, 4, 3]$$

If every vertex has the same degree, the graph is called **regular**. In a **k-regular** graph each vertex has degree k. As a special case, 3-regular graphs are also called **cubic graphs**.

When considering degree sequences, it is common practice to focus only on simple graphs, that is, graphs without loops and multiple edges. An interesting question that comes to mind is when we are given a list of numbers, is there also a simple graph whose degree sequence corresponds to that list? There are some obvious cases where we already know that a given list cannot correspond to a degree sequence. For example, we have just proven that the sum of vertex degrees is always even. Therefore, a minimal requirement is that the sum of the elements of that list should be even as well. Likewise, it is not difficult to see that, for example, the sequence $[4, 4, 3, 3]$ cannot correspond to a degree sequence. In this case, if this were a degree sequence, we would be dealing with a graph of four vertices. The first vertex is supposed to have four incident edges. In the case of simple graphs, each of these edges should be incident with a different vertex. However, there are only three vertices left to choose from, so $[4, 4, 3, 3]$ can never correspond to the degree sequence of a simple graph.

Of course, taking a trial-and-error approach to see whether a list corresponds to a degree sequence is not the way to go. Fortunately, there is a systematic way to see whether a given list of numbers corresponds to the degree sequence of a simple graph, in which case the sequence is said to be **graphic**. Let's return to our graph from Figure 2.1, but now assume that we are given only the list $[7, 5, 5, 4, 4, 4, 4, 3]$. We ask ourselves whether this list is graphic. If this is the case, we should be able to construct a graph that has this degree sequence. Note that this graph need not necessarily be the same as the one from Figure 2.1. This is how we can address this issue.

- Consider $[7, 5, 5, 4, 4, 4, 4, 3]$. If this sequence is graphic corresponding to a graph, say G_1, then we should be able to construct G_1 from another graph G_2 by adding a vertex v_1 to G_2 and joining v_1 to *seven* other vertices from G_2. This would then explain that G_1 has a vertex with highest degree 7. Note that for this construction to work, it is necessary that we can construct G_2.

    It should be clear that if we do not change the ordering of vertex degrees, that the degree sequence of G_2 is equal to $[4, 4, 3, 3, 3, 3, 2]$. First, it contains one element less than the degree sequence of G_1. Second, the first element of the degree sequence of G_2 corresponds to the second element of G_1's degree sequence: it's the degree of the same vertex, yet for G_2 it should be one less than in G_1 because this vertex is not yet joined to the added vertex v_1. Likewise, the second element of G_2's degree sequence corresponds to the third one in the degree sequence of G_1, and so on.

- If $[4, 4, 3, 3, 3, 3, 2]$ is graphic we can apply the same trick: G_2 should be constructable from a graph G_3 by adding a vertex v_2 and joining v_2 to *four* vertices from G_3. Following a completely analogous procedure as before, v_2 is joined to the vertices from G_3 such that these vertices will then have vertex degree 4, 3, 3, and 3, respectively. This can only mean that in G_3 they will have degree 3, 2, 2, and 2, respectively, leading to the following list: $[3, 2, 2, 2, 3, 2]$.

    Note that in this example, the fifth element is the same as the sixth element in the degree sequence of G_2. The first four elements represent vertices that will be joined to the new vertex v_2. The other elements represent vertices that remain untouched, and will thus have the same number of incident edges in G_2.

- Continuing this line of reasoning, if $[3, 3, 2, 2, 2, 2]$ is the (now ordered) degree sequence of G_3, then we should be able to construct G_3 from a graph G_4 to which we have added a vertex v_3. This vertex would be joined to the vertices having degree 2, 1, and 1 in G_4, respectively, yielding the list $[2, 1, 1, 2, 2]$. Again, note that this list contains one element less than the degree sequence of G_3, but that now its *fourth* and subsequent elements represent vertices that have the same vertex degree in G_4 and G_3.

- We now have that if ordered list $[2, 2, 2, 1, 1]$ is graphic, then so should $[1, 1, 1, 1]$, corresponding to a graph G_5.

- Likewise, if $[1, 1, 1, 1]$ is graphic, then so should the list of vertex degrees $[0, 1, 1]$ correspond to a graph G_6.

- Finally, if the ordered list $[1, 1, 0]$ is graphic, then so should $[0, 0]$,

2.1. FORMALITIES

which is true: it is a graph G_7 with two vertices and no edges.

We can safely conclude that the sequence $[7, 5, 5, 4, 4, 4, 4, 3]$ indeed corresponds to a simple graph. The construction of the graph G_1 is illustrated in Figure 2.2 which shows how each graph $G_1, G_2, \ldots, G_6$ is constructed by adding a vertex to the previous one, starting from graph G_7. The answer to whether G_1 is the same as the graph from Figure 2.1 is a question we defer until later. In fact, it turns out to be question that is generally not easy to resolve.

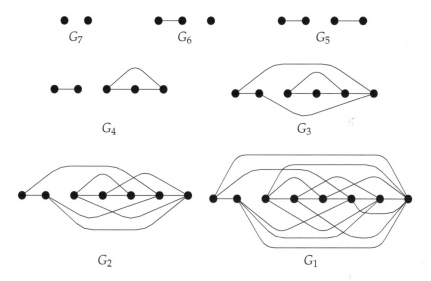

Figure 2.2: The construction of graph G_1 from previous graphs based on degree sequences.

Intuitively, it should be clear that we have just introduced a *systematic* way of checking whether a given list of numbers corresponds to the degree sequence of a graph. It also forms the essence of the proof of the following theorem that tells us when a list of numbers is indeed graphic.

Theorem 2.2 (Havel-Hakimi): *Consider a list* $s = [d_1, d_2, \ldots, d_n]$ *of n numbers in descending order. This list is graphic if and only if* $s^* = [d_1^*, d_2^*, \ldots, d_{n-1}^*]$ *of* $n - 1$ *numbers is graphic as well, where*

$$d_i^* = \begin{cases} d_{i+1} - 1 & \text{for } i = 1, 2, \ldots, d_1 \\ d_{i+1} & \text{otherwise} \end{cases}$$

> **Note 2.3** (Mathematical language)
> Note that this theorem consists of *two* statements:
>
> 1. if **s*** is graphic then so is **s**
> 2. if **s** is graphic then so is **s***
>
> This is the meaning of "if and only if," which is often abbreviated to *iff*. We will encounter more of such theorems, and in order to prove them correct, proofs in these cases will always consist of two parts.

Proof of Theorem 2.2. To prove this theorem, let us first assume that **s*** is graphic. We then need to show that **s** is also graphic. Let G^* be a simple graph with degree sequence **s***. We now construct a simple graph G from G^* with degree sequence **s** as follows (and in doing so, we show that **s** is graphic). Take G^* and add a vertex u. For readability, let $k = d_1$ and consider the k vertices $v_1, v_2, \ldots, v_k$ from G^* having respectively degree $d_1^*, d_2^*, \ldots, d_k^*$. We then join these vertices to the newly added vertex u. Obviously, u now has degree k, but also each vertex v_i now has degree $d_i^* + 1$. Because all other vertices of G^* are not joined with u, their vertex degree is left unaffected. As a consequence, the newly constructed graph G has degree sequence $[k, d_1^* + 1, d_2^* + 1, \ldots, d_k^* + 1, d_{k+1}^*, \ldots, d_{n-1}^*]$, which is precisely **s**.

Let us now consider the opposite: if **s** is graphic, we need to show that **s*** is so as well. In other words, we need to find a graph G^* that has degree sequence **s***. To this end, we consider three different sets of vertices from G. Let u be a vertex with degree $k = d_1$. Let $V = \{v_1, v_2, \ldots, v_k\}$ be the respective vertices with the k next highest degrees $d_2, d_3, \ldots, d_{k+1}$. Finally, let $W = \{w_1, w_2, \ldots, w_{n-k-1}\}$ be the remaining $n - k - 1$ vertices with degree $d_{k+2}, d_{k+3}, \ldots, d_n$, respectively.

Consider the graph G^* by removing u from G, along with the k edges incident with u. If each of these edges is incident with one of the vertices from V, then obviously G^* is a graph with degree sequence $(d_2 - 1, d_3 - 1, \ldots, d_{k+1} - 1, d_{k+2}, \ldots, d_n)$, which is precisely **s***.

Now consider the situation that u is adjacent to a vertex from W, say w_i. If for some vertex $v_j \in V$, the degree of v_j and w_i are the same, i.e., $\delta(w_i) = \delta(v_j)$, then we can simply swap w_i and v_j in the original construction of the sets V and W, meaning that $\langle u, w_i \rangle$ is now an edge incident with a vertex from V instead of W. However, if $\delta(w_i) < \delta(v_j)$ (i.e., $\delta(w_i)$ is less than the degree of any vertex from V) we cannot apply such an exchange.

The problem that we need to solve is that there is now a vertex v_j not adjacent to u whose degree will remain the same instead of being decremented by 1. Likewise, by simply removing u we would decrease the degree of w_i,

2.1. FORMALITIES

while we would like to see it unaffected if we want to realize the degree sequence $\mathbf{s}^*$. Note, however, that because $\delta(v_j) > \delta(w_i)$, there is a vertex x adjacent to v_j but not adjacent to w_i (note also that $x \neq u$), as shown in Figure 2.3(a). In constructing G^* we now first remove edges $\langle u, w_i \rangle$ and $\langle v_j, x \rangle$, and then add edges $\langle x, w_i \rangle$ and $\langle u, v_j \rangle$, leading to the situation shown in Figure 2.3(b). The effect is that we now have a graph G' in which u is adjacent to v_j instead of w_i, but without affecting the degree of u, v_j, x, or w_i. In other words, G' has the degree sequence $\mathbf{s}$. If u is now adjacent to vertices only from V, we have already shown that $\mathbf{s}^*$ is graphic. If u is still adjacent to a vertex from W, we apply the same method to construct a graph G'' in which u is adjacent to one more vertex from V. If necessary, we repeat this method until u is adjacent only to vertices from V, at which point we know that $\mathbf{s}^*$ is graphic. □

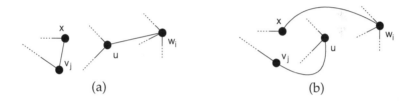

Figure 2.3: Changing a graph so that it meets the sets V and W of the Havel-Hakimi proof.

Note 2.4 (Proof techniques)
The proof of the Havel-Hakimi theorem illustrates a number of important issues in graph theory. In the first place, it is a **proof by construction**. In the case of the Havel-Hakimi theorem this means that we show that the theorem holds by actually constructing a graph from a given degree sequence. In general, proving properties by construction is very powerful: not only do we demonstrate the *existence* of a property, we also show how to get there. In contrast, with *nonconstructive proofs* we merely prove that some property must exist, often by first assuming that it does not exist and subsequently arriving at a contradiction. We will come across more of these proofs, but also ones in which we merely show that a property must exist, without giving a graph that has the specific property.

Another important issue in proving the Havel-Hakimi theorem, is that we show the power of visualization. Visualizing situations, either explicitly on paper or otherwise merely in your mind, is particularly useful in the case of graphs, and should come as no surprise. When graphs are studied for the first time, it is tempting to draw complete examples, that is, graphs in which each

> edge joins two vertices. However, as you become more experienced, it turns out that sketching graphs as is done in Figure 2.3 is actually more illustrative as these drawings reflect the essence of what you are trying to prove. Irrelevant details are thus avoided. You are encouraged to go for the sketches.

Note that two graphs with the same degree sequence need not be the same. In other words, when given a degree sequence, it may be possible to construct several, different, graphs that have that sequence, as is illustrated in Figure 2.4. The two graphs in Figure 2.4(a) have the same degree sequence, yet they are truly different. The same holds for the two graphs from Figure 2.4(b). We return to the notion of similarity of graphs in Section 2.2.

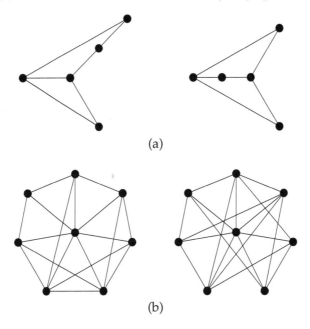

Figure 2.4: Different graphs with the same ordered degree sequence: (a) [3, 3, 2, 2, 2], and (b) [7, 5, 5, 4, 4, 4, 4, 3].

Subgraphs and line graphs

Another important concept of graphs is that of a subgraph. A graph H is a subgraph of G if H consists of a subset of the edges and vertices of G, such that the end points of edges in H are also contained in H. Strictly speaking, we have the following:

2.1. FORMALITIES

Definition 2.4: *A graph H is a **subgraph** of G if $V(H) \subseteq V(G)$ and $E(H) \subseteq E(G)$ such that for all $e \in E(H)$ with $e = \langle u,v \rangle$, we have that $u,v \in V(H)$. When H is a subgraph of G, we write $H \subseteq G$.*

As an example, Figure 2.5 shows a so-called **cubic graph** (i.e., 3-regular graph) with 8 vertices and three of its subgraphs.

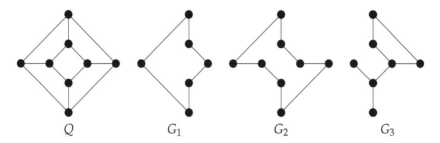

Figure 2.5: The cubic graph Q with 8 vertices and three subgraphs G_1, G_2, and G_3.

When analyzing properties of graphs, it is often convenient to consider subgraphs formed by a specific subset of vertices. These are so-called induced subgraphs, which are constructed by taking a subset V^* of vertices and adding each edge from the original graph that connects two vertices from V^*. Formally, we have:

Definition 2.5: *Consider a graph G and a subset $V^* \subseteq V(G)$. The **subgraph induced by V^*** has vertex set V^* and edge set E^* defined by*

$$E^* \stackrel{\text{def}}{=} \{e \in E(G) | e = \langle u,v \rangle \text{ with } u,v \in V^*\}$$

Likewise, if $E^* \subseteq E(G)$, the subgraph induced by E^* has edge set E^* and a vertex set V^* defined by

$$V^* \stackrel{\text{def}}{=} \{u,v \in V(G) | \exists e \in E^* : e = \langle u,v \rangle\}$$

The subgraph induced by V^* or E^* is written as $G[V^*]$ or $G[E^*]$, respectively.

Clearly, every simple graph $G = (V, E)$ having n vertices can be seen as a subgraph of the complete graph K_n. Moreover, if we consider its complement $\overline{G} = (V, \overline{E})$, then the **union** of G and $\overline{G}$, that is, the graph with vertex set V and edge set $E \cup \overline{E}$, corresponds to K_n. This is what we have previously coined taking two graphs "together."

Somewhat related to the notion of an induced subgraph is that of a line graph.

Definition 2.6: *Consider a simple graph $G = (V, E)$. The **line graph** of G, denoted as $L(G)$ is constructed from G by representing each edge $e = \langle u, v \rangle$ from E by a vertex v_e in $L(G)$, and joining two vertices v_e and v_{e^*} if and only if edges e and e^* are incident with the same vertex in G.*

To illustrate, consider the graph shown in Figure 2.6(a), containing four vertices and six edges. Its line graph, shown in Figure 2.6(b), consists of six vertices.

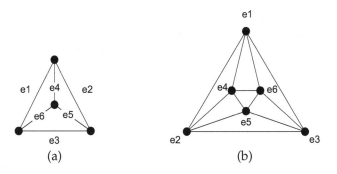

Figure 2.6: (a) A graph G and (b) its line graph $L(G)$.

Note 2.5 (Mathematical language)
Note that we used one of those awkward, yet precise mathematical statements when defining a subgraph induced by a set of edges. In this case, the mathematical statement

$$V^* \stackrel{\text{def}}{=} \{u, v \in V(G) | \exists e \in E^* : e = \langle u, v \rangle\}$$

should be translated into plain English as follows:

> V^* is the set of vertices from $V(G)$ formed by the end points of edges in E^*.

If we would literally translate from math, we would have

> V^* is defined by all vertices u and v from $V(G)$ for which there exists an edge in E^* that joins u and v.

When reading this second version, it is important to try to move away from all the math and come up with something like the first one, which is more intuitive and actually simpler.

A special induced subgraph is the one by which we simply remove a specific vertex, say v: $G[V(G) \backslash \{v\}]$. We came across this type of graph in our proof

2.2 Graph representations

of Theorem 2.2. Instead of using the notation $G[V(G)\setminus\{v\}]$ we will often simply write $G - v$. Likewise, if e is an edge, we will often write $G - e$ instead of $G[E(G)\setminus\{e\}]$. Similar simplified notations will be used when dealing with subsets of vertices or edges, respectively.

2.2 Graph representations

It should be clear from the presentation so far that graphs can be drawn in different ways, but also that when considering their formal definition, they are merely described in terms of vertices and edges. Let us now pay attention to how we can conveniently represent graphs. This issue is particularly important when we need to represent very large graphs for automated processing by computers.

Data structures

There are different ways to represent graphs. Perhaps the most appealing one is to use an **adjacency matrix**. Consider a graph G with n vertices and m edges. Its adjacency matrix is nothing else but a table $\mathbf{A}$ with n rows and n columns with entry $\mathbf{A}[i, j]$ denoting the number of edges joining vertex v_i and v_j. To illustrate, Figure 2.7 shows a simple graph with its accompanying adjacency matrix.

It is not difficult to see that the following properties hold:

- An adjacency matrix is *symmetric*, that is for all i, j, $\mathbf{A}[i,j] = \mathbf{A}[j,i]$. This property reflects the fact that an edge is represented as an *unordered* pair of vertices $e = \langle v_i, v_j \rangle = \langle v_j, v_i \rangle$.

- A graph G is simple if and only if for all i, j, $\mathbf{A}[i,j] \leq 1$ and $\mathbf{A}[i,i] = 0$. In other words, there can be at most one edge joining vertices v_i and v_j and, in particular, no edge joining a vertex to itself.

- The sum of values in row i is equal to the degree of vertex v_i, that is, $\delta(v_i) = \sum_{j=1}^{n} \mathbf{A}[i,j]$.

As an alternative, we can also use an **incidence matrix** of a graph as its representation. An incidence matrix $\mathbf{M}$ of graph G consists of n rows and m columns such that $\mathbf{M}[i,j]$ counts the number of times that edge e_j is incident with vertex v_i. Note that $\mathbf{M}[i,j]$ is either 0, 1, or 2: an edge can be only *not* incident with vertex v_i, it has vertex v_i as exactly one of its end points, or is a loop joining vertex v_i with itself. Figure 2.8 shows the incidence matrix for the graph from Figure 2.7. Again, the following properties are easy to verify:

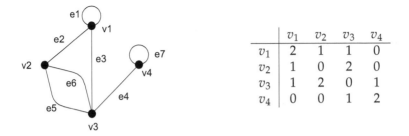

Figure 2.7: A graph with its associated adjacency matrix.

- A graph G has no loops if and only if for all i, j, $\mathbf{M}[i,j] \leq 1$.
- The sum of all values in row i is equal to the degree of vertex v_i. In mathematical terms, this is expressed as $\forall i : \delta(v_i) = \sum_{j=1}^{m} \mathbf{M}[i,j]$.
- Because each edge has exactly two, not necessarily distinct end points, we know that for all j, $\sum_{i=1}^{n} \mathbf{M}[i,j] = 2$.

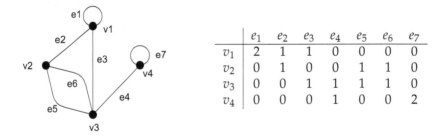

Figure 2.8: A graph with its associated incidence matrix.

One of the problems with using either an adjacency matrix or an incidence matrix is that without further optimizations, the total number of elements for representing a graph is $n \times n$ or $n \times m$, respectively. This is not very efficient when having to deal with very large graphs, especially when the number of edges is relatively small. To see why this is true, consider the representation of an adjacency matrix in a computer. Assume that we use only a single byte to count the number of edges joining a pair of vertices. Without any further optimizations, a graph with 100,000 vertices would require a total of 100,000 × 100,000 bytes of storage, that is, close to 10 Gbyte. Using an incidence matrix and assuming a total of 250,000 edges, a straightforward, nonoptimized representation would require close to 25 Gbytes of

2.2. GRAPH REPRESENTATIONS

storage. Both representations, even when applying all kinds of storage optimizations, generally tend to be rather inefficient.

An often more efficient representation, and used in practice, is that of an **edge list**. In this case, we merely list the edges of a graph G by specifying for each edge which vertices it is incident with. Note that this representation grows linearly with the number of edges. For example, the edge-list representation of the graph from Figure 2.8 is:

$$(\langle v_1, v_1\rangle, \langle v_1, v_2\rangle, \langle v_1, v_3\rangle, \langle v_2, v_3\rangle, \langle v_2, v_3\rangle, \langle v_3, v_4\rangle, \langle v_4, v_4\rangle)$$

In particular, with m edges, we would need to store only $2 \cdot m$ data items. Assuming that a vertex can be represented by four bytes, this means that for our example graph with 100,000 vertices and 250,000 edges, we would need only close to 2 Mbytes of storage. In practice, this number will be larger because we need additional data structures to easily navigate through the edge list. Nevertheless, the total amount of required storage will generally stay significantly less than what is needed for an adjacency or incidence matrix.

It should be clear that by simply going through this list, we also find the vertices of the associated graph, provided that each vertex is incident with at least one edge. In practice, an edge list is often accompanied by a list of vertices, for example, to describe attached labels (such as "v1").

Graph isomorphism

An important observation is that all these representations are independent of the way that we draw a graph. Consider the graphs shown in Figure 2.9. No matter whether we represent each graph by its adjacency matrix, incidence matrix, or edge list, if we properly attach labels to vertices and edges, we will find that their respective representations are exactly the same. As a consequence, they should also be considered to be the same. This notion of similarity is formalized through what is known as graph isomorphism.

Definition 2.7: *Consider two graphs $G = (V, E)$ and $G^* = (V^*, E^*)$. G and G^* are **isomorphic** if there exists a one-to-one mapping $\phi : V \to V^*$ such that for every edge $e \in E$ with $e = \langle u, v \rangle$, there is a unique edge $e^* \in E^*$ with $e^* = \langle \phi(u), \phi(v) \rangle$.*

Stated differently, two graphs G and G^* are isomorphic if we can uniquely map the vertices and edges of G to those of G^* such that if two vertices were joined in G by a number of edges, their counterparts in G^* will be joined by the same number of edges.

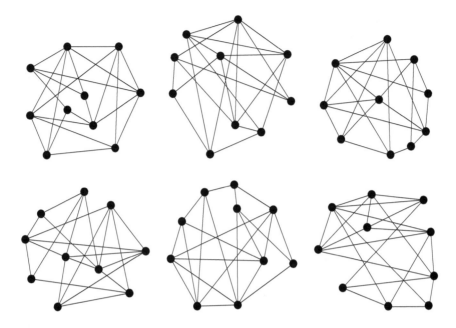

Figure 2.9: Six different drawings of graphs with the same representation, that is, isomorphic graphs.

Note 2.6 (Mathematical language)
Couldn't we just talk about the *same* graphs, you might wonder, instead of using a term like *isomorphism*? However, "isomorphism" is a well-defined mathematical concept that is used for more than just graphs. In essence, it is used in those situations where we are dealing with sets (like vertices), and that the elements in those sets are somehow organized in a specific way. Isomorphism is then used to express that two sets have essentially the same elements when you ignore labeling issues, but also that their organization is the same. An isomorphism is then a *structure-preserving* mapping between two sets.

In many cases, checking whether two graphs are isomorphic is relatively simple as there are a number of important *necessary* requirements that need to be fulfilled. For example, it should be obvious that the two graphs need to have the same number of vertices and edges in order to be isomorphic. A stronger requirement is that they have the same ordered degree sequence. This may same obvious, but if we want to be precise, showing the obvious may turn out to be more cumbersome than expected. Let's consider the following formal formulation.

2.2. GRAPH REPRESENTATIONS

Theorem 2.3: *If two graphs G and G^* are isomorphic, then their respective ordered degree sequences should be the same.*

Proof. Let ϕ be the one-to-one mapping by which G and G^* are known to be isomorphic. Consider vertex u from G and its adjacent vertices $v_1, \ldots, v_k$. By definition, each edge $e_i = \langle u, v_i \rangle$ incident with u in G is mapped to a unique edge $e_i^* = \langle \phi(u), \phi(v_i) \rangle$ in G^*. Because each edge e_i^* is incident with $\phi(u)$, we must have that $\delta(u) \leq \delta(\phi(u))$.

Now consider a vertex $v^* \in V(G^*)$ that is adjacent to $\phi(u)$. By definition of isomorphism, we know that the edge $e^* = \langle \phi(u), v^* \rangle$ must uniquely map to an edge $e = \langle \phi^{-1}(\phi(u)), \phi^{-1}(v^*) \rangle$ in G, where ϕ^{-1} denotes the *inverse* mapping of ϕ. Because ϕ is a one-to-one mapping, we also know that $\phi^{-1}(\phi(u)) = u$, and thus that $e = \langle u, \phi^{-1}(v^*) \rangle$. In other words, every edge incident with $\phi(u)$ in G^* will be incident with u in G. This means that $\delta(\phi(u)) \leq \delta(u)$.

We conclude that $\delta(u) = \delta(\phi(u))$ for all vertices of G, implying that the ordered degree sequences of G and G^* should be the same. □

Unfortunately, this theorem gives us only a *necessary* condition for two graphs to be isomorphic, yet it is not a *sufficient* condition. In other words, if two graphs have the same ordered degree sequence, then that fact alone is not sufficient to conclude that they are also isomorphic. Yet to be isomorphic, it is *necessary* for their respective ordered degree sequences to be the same.

Note 2.7 (Mathematical language)
The difference between *necessary* and *sufficient* conditions seems an obvious one, yet they are surprisingly often confused in mathematical proofs. Formally, in graph theory, conditions are used to prove properties of graphs. When a condition C is said to be *necessary*, this means that a property P can hold *only if* C is met. When a condition C is said to be *sufficient*, this means that *if* C is met, then property P will hold true. And indeed, when property P is true *if and only if* condition C is met, indicates that C is a *necessary* and *sufficient* condition for property P to be valid.

To illustrate, consider the graphs from Figure 2.4(a), which are shown again in Figure 2.10. Although they have the same ordered degree sequence, they are not isomorphic. One way of seeing this is that the two vertices with degree 3 are adjacent to one another in G, but not in G^*. (There are other structural differences, yet explaining these requires the introduction of more graph concepts, which we defer until later.)

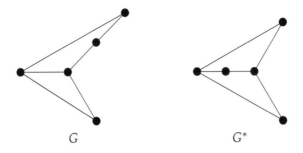

Figure 2.10: Two nonisomorphic graphs with the same ordered degree sequence.

The bad news is that there are no known easy sufficient conditions that will tell us in general whether two graphs are isomorphic or not. Essentially, this means that once we have found that all necessary conditions have been fulfilled, we will have to resort to a trial-and-error method. For example, with the graphs from Figure 2.10, we were able to successfully consider whether the highest-degree vertices were adjacent in both graphs. In other cases, however, we may have to look at other properties.

Note 2.8 (More information)
In the worst case, we may have to resort to an *exhaustive* method. Consider a graph G with n vertices $\{v_1, v_2, \ldots, v_n\}$, and a graph G^* also with n vertices. To check for isomorphism, we need to find a one-to-one mapping between these two vertex sets. With an exhaustive approach, we simply go through all possible mappings to see if there is one that establishes isomorphism. Unfortunately, there may be quite a few mappings that we need to check. To be precise, there are potentially $n!$ mappings to consider, where

$$n! \stackrel{\text{def}}{=} n \cdot (n-1) \cdot (n-2) \cdots 2 \cdot 1$$

(to be pronounced as n *factorial*). This is relatively easy to see as follows. For any mapping, we have n choices for mapping v_1 to one of the vertices from G^*. After that, there are $n-1$ possibilities left for mapping v_2 to a vertex from G^*, and then another $n-2$ for mapping v_3, and so on. Finally, after having made a choice for each vertex $v_1, v_2, \ldots, v_{n-1}$, we have only one more option left for v_n.

Checking $n!$ mappings is no pleasure game—consider the following table:

n	$n!$	n	$n!$	n	$n!$
1	1	6	720	11	39,916,800
2	2	7	5040	12	479,001,600
3	6	8	40,320	13	6,227,020,800
4	24	9	362,880	14	87,178,291,200
5	120	10	3,628,800	15	1,307,674,368,000

> In fact, for large n, its factorial can be approximated by
>
> $$n! \approx \sqrt{2\pi n}(\frac{n}{e})^n$$
>
> which reaches amazingly high numbers even for relatively small values of n. There is also no chance that brute-force computations with a computer are going to bring any serious help here. For example, if a computer were able to check whether one specific mapping could establish isomorphism between two graphs in only 1 nanosecond (which is 10^{-9} seconds), it would still take about 500 years to go through all possible mappings for two 25-vertex graphs. More cleverness is needed.
>
> We note that algorithms do exist that can efficiently test isomorphism for many graphs up to approximately 100 vertices, with perhaps the fastest one being nauty devised by McKay [1980]. Also, efficient algorithms exist for graphs for which the maximal vertex degree is known to be bound by a constant [Luks, 1982].

2.3 Connectivity

In all the graphs we have considered so far, each vertex v could be reached from any other vertex w in the sense that we could indicate a chain of adjacent vertices from v to w. In this section, we will take a closer look at this important concept of connectivity. We start with some basic terminology:

Definition 2.8: *Consider a graph G. A $(\mathbf{v_0}, \mathbf{v_k})$-walk in G is an alternating sequence $[v_0, e_1, v_1, e_2 \ldots v_{k-1}, e_k, v_k]$ of vertices and edges from G with $e_i = \langle v_{i-1}, v_i \rangle$. In a closed walk, $v_0 = v_k$. A trail is a walk in which all edges are distinct; a path is a trail in which also all vertices are distinct. A cycle is a closed trail in which all vertices except v_0 and v_k are distinct.*

Using the notion of a path, we define a graph to be connected when there is a path between each pair of distinct vertices. Formally, we have:

Definition 2.9: *Two distinct vertices u and v in graph G are connected if there exists a (u,v) − path in G. G is connected if all pairs of distinct vertices are connected.*

Clearly, all the graphs we have considered so far are indeed connected. However, there is no reason to assume that a graph is always connected. If we take a look at the definition of a graph, there is nothing there that states that all vertices should be connected. Intuitively, this means that a graph could also consist as a collection of components, where each component is a connected subgraph. This can be made precise as follows:

Definition 2.10: *A subgraph H of G is called a **component** of G if H is connected and not contained in a connected subgraph of G with more vertices or edges. The number of components of G is denoted as $\omega(G)$.*

Note that a component is not just a subgraph: it is a *maximal, connected* subgraph. In other words, if we would consider a subgraph H of a graph G and would find that there is a vertex not in H that is connected to a vertex in H, then H is, by definition, not a component. Maximality also incorporates edges, meaning that if an edge e joins two vertices in G, e should be contained in H.

The notion of connectivity is important, notably when considering the *robustness* of networks. Robustness in this context means how well the network stays connected when we remove vertices or edges. For example, as we mentioned in Chapter 1, the Internet can be viewed as a (huge) graph in which routers form the vertices and communication links between routers the edges. In a formal sense, the Internet is connected. However, if it were possible to partition the network into multiple components by removing only a single vertex (i.e., router) or edge (i.e., communication link), we could hardly claim the Internet to be robust. In fact, it is extremely important for networks such as the Internet to be able to sustain serious attacks and failures by which routers and links are brought down, such that connectivity is still guaranteed.

There are many networks for which robustness in one way or another plays an important role. Let us now formalize this notion by considering what are known as vertex and edge cuts.

Definition 2.11: *For a graph G let $V^* \subset V(G)$ and $E^* \subset E(G)$. V^* is called a **vertex cut** if $\omega(G - V^*) > \omega(G)$. If V^* consists of a single vertex v, then v is called a **cut vertex**. Likewise, if $\omega(G - E^*) > \omega(G)$ then E^* is called an **edge cut**. If E^* consists of only a single edge e, then e is known as a **cut edge**.*

Note that we have used the notation $G - V^*$ to indicate the induced subgraph $G[V(G)\backslash V^*]$. What the definition states is that V^* is a vertex cut of a connected graph if the removal of vertices in V^* from G will make G disintegrate into several components. In other words, G will become disconnected. Analogously, an edge cut of G is a collection of edges that will make G fall apart into multiple components when those edges are removed. In the definition given above, we have used the simpler notation $G - E^*$ to indicate the induced subgraph $G[E(G)\backslash E^*]$.

Of particular interest is the *minimal* vertex cut for a connected graph. In other words, how many vertices do we need to remove from a connected graph before it becomes disconnected? An important observation is the following. Let $\kappa(G)$ denote the size of a minimal vertex cut for graph G, and

2.3. CONNECTIVITY

likewise, $\lambda(G)$ the size of a minimal edge cut. As it turns out, $\kappa(G) \leq \lambda(G)$, but also that $\lambda(G)$ is less or equal to the minimal vertex degree. Using the notation min S to denote the smallest value found among the elements in set S, these properties are formulated in the following important theorem.

Theorem 2.4: $\kappa(G) \leq \lambda(G) \leq \min\{\delta(v) | v \in V(G)\}$.

Proof. That $\lambda(G) \leq \min\{\delta(v) | v \in V(G)\}$ is easy to see. Consider a vertex u with minimal degree, that is, $\delta(u) = \min\{\delta(v) | v \in V(G)\}$. If we simply remove the $\delta(u)$ edges incident with u, then u will become isolated, and certainly the resulting graph will have at least one more component then it had before (namely the one consisting only of u).

To prove that $\kappa(G) \leq \lambda(G)$, consider a graph G with $\lambda(G) = k$ and let $E^* = \{e_1, e_2, \ldots, e_k\}$ be a minimal edge cut of G, with $e_i = \langle u_i, v_i \rangle$. Let U denote the set of vertices $\{u_1, \ldots, u_k\}$ and V the set $\{v_1, \ldots, v_k\}$. Note that in this case, the vertices in either set need not be distinct. The graph $G - E^*$ will fall apart into exactly two components, say G_1 and G_2 (we leave it to you to show that this is indeed true). If G_1 contains a vertex u distinct from any u_i, as shown in Figure 2.11(a), then clearly removing all vertices in U will disconnect u from any vertex in G_2, so that $\kappa(G) \leq k$.

If there is no such vertex u, then assume that $V(G_1) = U$. Consider vertex u_1. We know that u_1 is adjacent to d_1 vertices from G_1, and each of these neighbors in G_1 is adjacent to a vertex from V. Let E_1^* be a set of edges from E^* joining vertices from the d_1 neighbors of u_1 and exactly one vertex from V. Likewise, let E_2^* be the d_2 edges from E^* incident with u_1. This situation is shown in Figure 2.11(b). Obviously, $d_1 + d_2 = |E_1^* \cup E_2^*| \leq |E^*|$. Also, the $d_1 + d_2$ neighboring vertices of u_1 form a vertex cut, shown as open circles in Figure 2.11(b). This also means that $\kappa(G) \leq d_1 + d_2 \leq |E^*| = \lambda(G)$, completing the proof. □

A graph G for which $\kappa(G) \geq k$ for some k is said to be **k-connected**. Likewise, graph G is **k-edge-connected** if $\lambda(G) \geq k$. Finally, a graph for which $\kappa(G) = \lambda(G) = \min\{\delta(v) | v \in V(G)\}$ is said to be **optimally connected**.

Note 2.9 (Study tip)
The previous proof, and notably proving that $\kappa(G) \leq \lambda(G)$, is a typical example where graph theory requires insight. The proof is not obvious, and it can certainly not be expected that an undergraduate student would be able to devise it from scratch. What is important, however, is that the proof itself is understood well. To this end, you are encouraged to start with *reproducing* proofs, as this will enforce you to carefully think about every step that is taken. Simply being

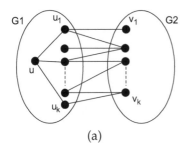 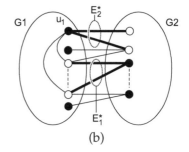

Figure 2.11: The two scenarios for the proof of Theorem 2.4.

> able to reproduce proofs is a well-known technique to successfully study graph theory.

What Theorem 2.4 tells us is that every graph is at most δ_{min}-edge connected, and at most δ_{min}-connected, where $\delta_{min} = \min\{\delta(v)|v \in V(G)\}$. We showed this for edge connectivity. Vertex connectivity is also easy to see: simply remove the δ_{min} vertices adjacent to a vertex of degree δ_{min} and the latter becomes disconnected. Of course, finding a lower bound for k is more interesting, but this turns out to be a relatively difficult problem to solve. Without going into the rather intricate details, we can say something about a lower bound for k by considering the notion of path independence.

Definition 2.12: *Consider a graph G and a collection* **P** *of (u,v)-paths in G, with $u,v \in V(G)$.* **P** *is* ***vertex independent*** *if for all (u,v)-paths $P_1, P_2 \in$* **P** *we have that $V(P_1) \cap V(P_2) = \{u,v\}$. The collection is* ***edge independent*** *if for all its (u,v)-paths P_1 and P_2, we have that $E(P_1) \cap E(P_2) = \emptyset$.*

In other words, two (u,v)-paths P_1 and P_2 are vertex independent if they share only the vertices u and v, and are edge independent if they have no edge in common. Using path independence, we now come to one of the more fundamental theorems in graph theory, formulated by the Austrian mathematician Karl Menger.

Theorem 2.5 (Menger): *Let G be a connected graph and u and v two nonadjacent vertices in G. The minimum number of vertices in a vertex cut that disconnects u and v is equal to the maximum number of pairwise vertex-independent paths between u to v. Analogously, the minimum number of edges in an edge cut that disconnects u and v, is equal to the maximum number of pairwise edge-independent paths between u and v.*

2.3. CONNECTIVITY

We omit the proof, and instead refer the interested reader to Bondy and Murty [2008], Diestel [2005], or West [2001].

> **Note 2.10** (Mathematical language)
> Menger's theorem should be read carefully: it mentions *pairwise independent paths*. In this case, the adjective *pairwise* is used to make clear that we should always consider pairs of paths when considering independence. And indeed, this makes sense when you would consider trying to count the number of *independent paths*: being an independent path can only be relative to another path.
> To complete the story, also note that the theorem is all about counting the number of (u,v)-paths, and not the number of *pairs* of such paths. In other words, *pairwise* is an adjective to *independent*, and not to *paths*.

It is not difficult to see that Menger's theorem leads to the following corollary:

Corollary 2.2: *A graph G is k-connected if and only if any two distinct vertices are connected by at least k pairwise vertex-independent paths. G is k-edge connected if and only if any two distinct vertices are connected by at least k pairwise edge-independent paths.*

Of particular interest is the following one:

Corollary 2.3: *Each edge of a 2-edge-connected graph lies on a cycle.*

This corollary actually follows from the previous one, which states (for the special case $k = 2$) that a graph is 2-edge-connected if and only if any two distinct vertices are connected by at least 2 pairwise edge-independent paths. The latter, of course, together form a cycle. We will use this corollary in the next chapter when discussing so-called directed graphs.

Intuitively, it should be clear that for any simple graph G a higher value of $\kappa(G)$, i.e., the size of a minimal vertex cut, implies that more edges are needed. We have just seen that in every k-connected graph each vertex will have at least k incident edges. Knowing that $\sum \delta(v) = 2 \cdot m$, this means that for a graph with n vertices, we would need $\frac{1}{2} \sum \delta(v)$ and thus at least $\frac{1}{2} \sum k = \frac{1}{2} n \cdot k$ edges. But what is the minimal number of edges for a graph to be k-connected? This question brings us to a so-called Harary graph:

Definition 2.13: *A **Harary graph** $H_{k,n}$ is a k-connected simple graph with n vertices and with a minimal number of edges.*

What we now need to figure out is actually how many edges an Harary graph has. We will show that $H_{k,n}$ has exactly $\lceil k \cdot n/2 \rceil$ edges, that is, the smallest natural number of edges greater or equal to $k \cdot n/2$. To this end, we label the vertices in $H_{k,n}$ as $\{0, 1, \ldots, n-1\}$ and organize them graphically as a circle. Following Bondy and Murty [1976], we consider the following three cases for combinations of k and n.

k is even: We construct $H_{k,n}$ by joining each vertex i to its $k/2$ closest left-hand (i.e., clockwise) neighbors and its $k/2$ closest right-hand (i.e., counterclockwise) neighbors[1].

k is odd, n is even: In this case, we construct $H_{k-1,n}$ and add $n/2$ edges by joining vertex i to its left-hand neighbor at distance $\frac{n}{2}$ (with $0 \leq i < \frac{n}{2}$). In other words, we add edges $\langle 0, \frac{n}{2} \rangle, \langle 1, 1 + \frac{n}{2} \rangle, \langle 2, 2 + \frac{n}{2} \rangle, \ldots, \langle \frac{n-2}{2}, n-1 \rangle$.

k is odd, n is odd: In this case, we again first construct $H_{k-1,n}$ and then add the $(n+1)/2$ edges $\langle 0, \frac{n-1}{2} \rangle, \langle 1, 1 + \frac{n-1}{2} \rangle, \ldots, \langle \frac{n-1}{2}, n-1 \rangle$.

To clarify the construction of these graphs, Figure 2.12 shows graphs $H_{4,8}$, $H_{5,8}$, and the construction of $H_{5,9}$ from $H_{4,9}$.

Note 2.11 (More information)
At first sight, constructing Harary graphs seems to be one of those typical mathematical topics: nice, but it looks as if someone got carried away a bit. In fact, Harary graphs address a very relevant question in communication networks: trading off the costs between reliability and the number of communication links. A communication network constructed as an Harary graph $H_{k,n}$ tells us that we can remove up to k vertices before the network becomes partitioned. This means that if we are considering networks that are designed to disseminate data to every node, Harary graphs will give us the means to make them just as robust as we want them to be, yet with a minimal number of links. There are a number of variations on this theme, as explained by McQuillan [1977].

Admittedly, when first thought of, people considered the monetary costs of a communication link. With the robustness of the Internet, the problem seems to be less relevant. However, suppose we formulate costs in terms of how quickly data is disseminated. As we shall discuss in Chapter 8, we often want to construct an artificial, or virtual network on top of an existing communication network such as the Internet. In that case, we can shape the network as we like. As discussed by Jenkins and Demers [2001], Harary graphs are useful for constructing virtual networks that will optimally disseminate data in a group of nodes, while keeping that group k-connected.

Now that we have the procedure to construct Harary graphs, we need to show that they indeed have a minimal number of edges while maintaining the property of being k-connected. We first prove connectivity.

---

[1] Of course, being a left-hand or right-hand neighbor makes sense only if we assume that a vertex has an orientation. In our example, we orient a vertex toward the *middle* of the ring.

2.3. CONNECTIVITY

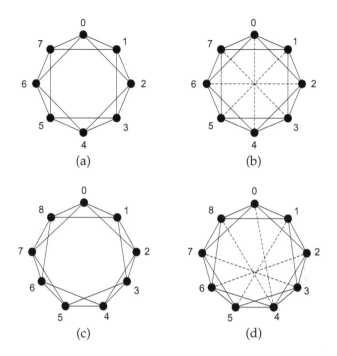

Figure 2.12: Various Harary graphs: (a) $H_{4,8}$, (b) $H_{5,8}$, (c) $H_{4,9}$, and (d) $H_{5,9}$. Dashed edges are the ones added to obtain $H_{5,8}$ from $H_{4,8}$, and $H_{5,9}$ from $H_{4,9}$, respectively.

Theorem 2.6: *The Harary graph $H_{k,n}$ is k-connected.*

Proof. Let us first consider the case that k is even. Our proof is completed if we can show that there is no vertex cut with fewer than k vertices. To this end, let us assume that such a set W *does* exist. If we can then prove that this assumption can never hold, we will have completed our proof (we come back to this method of proving a theorem below).

To this end, let vertices i and j belong to different components of $H_{k,n} - W$ (i.e., $G[V(H_{k,n})\setminus W]$). Consider the set $N_{i \to j}$ of left-hand neighbors of i, including i: $\{i, i+1, \ldots, j-1, j\}$, and likewise its right-hand neighbors $N_{i \leftarrow j} = \{j, j+1, \ldots, i-1, i\}$. In both cases, addition is taken modulo n. Let $W_{i \to j} \stackrel{\text{def}}{=} W \cap N_{i \to j}$ and $W_{i \leftarrow j} \stackrel{\text{def}}{=} W \cap N_{i \leftarrow j}$ (meaning that $W = W_{i \to j} \cup W_{i \leftarrow j}$). We know that $|W| < k$, so we must have that either $|W_{i \to j}| < k/2$ or $|W_{i \leftarrow j}| < k/2$, as is illustrated in Figure 2.13. Assume that $|W_{i \to j}| < k/2$

Now consider an arbitrary vertex u in $H_{k,n} - W$, lying on, say, segment S_1 (see Figure 2.13). We know that u is adjacent to $k/2$ consecutive vertices in either direction. As a consequence, removing less than $k/2$ vertices as is done through $W_{i \to j}$ will still allow us to reach any vertex v on segment S_2. In other words, $H_{k,n} - W$ will remain connected, contradicting our assumption that W was a vertex cut. □

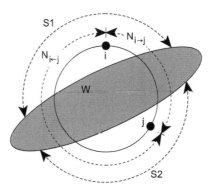

Figure 2.13: Illustration that $|W \cap N_{i \to j}| < k/2$ or $|W \cap N_{i \leftarrow j}| < k/2$.

Note 2.12 (Proof techniques)
We have just encountered our first **proof by contradiction**. This method is widely applied and you should definitely familiarize yourself with it. The principle is fairly straightforward: if you want to prove some statement P to be true, then in a proof by contradiction you *assume* that P cannot hold and subsequently show that this assumption will lead to something nonsensical. Nonsense can then only mean that your assumption was incorrect.

There is another important principle that surfaced in the previous proof, which is known as the **pigeonhole principle**. This principle simply states that if n items need to be spread over $m < n$ boxes, then there will be at least one box containing more than one item. How did we apply this principle? In our proof, we noted that the set W contained less than k elements and that we split it into two parts $W_{i \to j}$ and $W_{i \leftarrow j}$. The pigeonhole principle tells us that at least one of these two sets much have less than $k/2$ elements.

What remains is to show that a Harary graph also has a minimal number of edges. For any k-connected graph, we now know that each vertex v has a degree $\delta(v) \geq k$. Let $m_k(n)$ denote the minimal number of edges for any simple, k-connected graph G. Because $|E(G)| = \frac{1}{2} \sum_{v \in V(G)} \delta(v) \geq \frac{1}{2} \sum_{v \in V(G)} \delta_{min} \geq \frac{n \cdot k}{2}$, we know that $m_k(n) \geq \frac{n \cdot k}{2}$. It is not difficult to verify that $|E(H_{k,n})| = \frac{nk}{2}$, meaning that an Harary graph indeed has a minimal number of edges.

2.4 Drawing graphs

As the saying goes, a picture does often say more than a 1000 words. This certainly also holds for drawing graphs. We have already seen various examples of how the same graph can be drawn in different ways. As it turns out, this subject is so important that researchers have spent considerable effort on devising algorithms for drawing graphs. In this section we take a closer look at some of the results.

Graph embeddings

To illustrate why properly drawing graphs may be important, consider the graphs from Figure 2.14, which shows the **Petersen graph**, a particular 3-regular graph. Clearly, by just looking at these drawings, it is not obvious that we are dealing with the same graph (i.e., same in the sense of isomorphic). This instantly brings up the issue of what makes a good drawing of a graph. In this example, either Figure 2.14(a) or (b) is arguably the best one.

Formally, when drawing graphs we are considering so-called **graph embeddings**: a representation of a graph on a surface where vertices are associated with points on that surface. In practice, we always consider the two-dimensional plane, but note that embeddings in three dimensions are also possible.

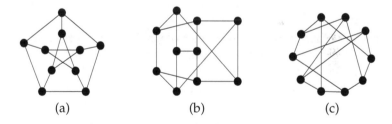

Figure 2.14: Three different drawings of the Petersen graph.

A commonly applied embedding is the **circular embedding**. In this case, the vertices are placed at evenly spaced points on a circle, as illustrated by the Petersen graph in Figure 2.14(c). The advantage of this representation is that no three vertices ever lie on the same straight line, or, in other words, are ever collinear. This has the effect that each edge can be easily drawn such that it remains visible in the drawing. This is an important property, notably when dealing with so-called **random graphs** in which pairs of vertices are connected by randomly chosen edges. In that case, it is generally important

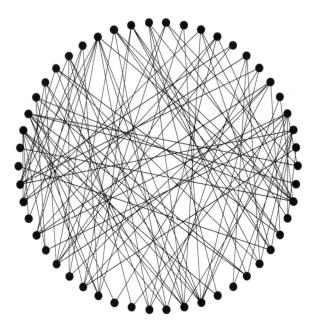

Figure 2.15: A random graph with 50 vertices and 103 edges. The circular embedding allows to draw each edge as a straight line that remains (reasonably) visible.

to see all edges. Figure 2.15 shows such a random graph with 50 vertices and 103 edges.

There are other useful embeddings to consider. For example, an important class is formed by **bipartite graphs**: graphs of which the set of vertices can be partitioned into two subsets such that no edge is incident to vertices from the same subset. In other words, each edge is incident to a vertex from either set:

Definition 2.14: *A graph G is **bipartite** if $V(G)$ can be partitioned into two disjoint subsets V_1 and V_2 such each edge $e \in E(G)$ has one end point in V_1 and the other in V_2: $E(G) \subseteq \{e = \langle u_1, u_2 \rangle | u_1 \in V_1, u_2 \in V_2\}$.*

Bipartite graphs are sometimes conveniently drawn as **ranked embeddings**. To explain, reconsider the graph from Figure 2.15. Although it is not obvious from the drawing at all, it turns out that this graph is actually bipartite. We can discover this by considering what the distance is between a given vertex v and each other vertex. The distance between two vertices v and w is informally expressed as the minimal number of edges between v and w (also called the **shortest path** to which we return in Chapter 4). This leads to a group of vertices at distance 1 (i.e., vertices adjacent to v,

2.4. DRAWING GRAPHS

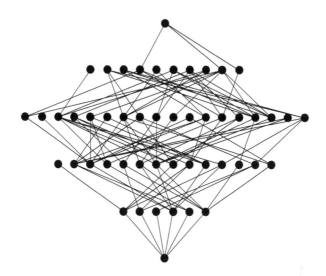

Figure 2.16: A ranked embedding of the graph from Figure 2.15.

which we had defined as its neighbors), at distance 2 (vertices adjacent to the vertices adjacent to v), etc. Now consider Figure 2.16 for which we have selected an arbitrary vertex v from G and subsequently (1) ranked all other vertices according to their distance, and (2) placed all vertices at the same distance along the same vertical line. What we observe is that there are no edges between vertices at the same distance. This can only mean that G is bipartite. In fact, in this example the set of vertices can be partitioned into six disjoint subsets.

These examples illustrate that examining graphs through visual inspection requires the use of computer tools. What these tools invariably do is compute vertex positions in the two-dimensional plane according to some simple or complex criterion. Circular and ranked embeddings are relatively simple. More complex ones involve spreading vertices far apart while still keeping connected ones close to each other. An example of such an approach is a **spring embedding** [Eades, 1984]. In this case, the vertices are modeled as rings connected by springs. Initially, the vertices are positioned randomly in the two-dimensional plane, after which the springs do their work by trying to reach an equilibrium. To illustrate, Figure 2.17 shows a number of steps by which the randomly positioned vertices in Figure 2.17(a) are gradually brought into an equilibrium.

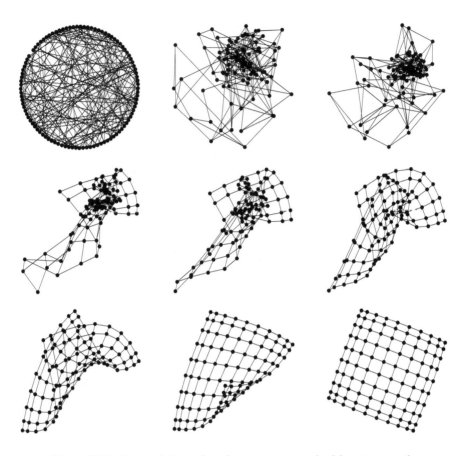

Figure 2.17: The evolution of applying a spring embedding to a graph.

Note 2.13 (More information)
As illustrated by Figure 2.17, spring embeddings can lead to an appealing visualization of a graph. Let's take a closer look at how the approach works. As stated previously, Eades [1984] proposed to represent each vertex as a ring and each edge as a spring. Each vertex u is initially positioned in a two-dimensional plane, with coordinates (u_x, u_y). Each spring $e = \langle u, v \rangle$ exerts an attracting force $F_{att}(u, v)$ between the vertices u and v it joins according to the formula

$$F_{att}(u,v) \stackrel{\text{def}}{=} \begin{cases} 2\log(d(u,v)) & \text{if } u \text{ and } v \text{ are adjacent} \\ 0 & \text{otherwise} \end{cases}$$

where $d(u, v)$ is the length of the spring between u and v. This length corresponds to the distance between u and v defined as:

$$d(u,v) \stackrel{\text{def}}{=} \sqrt{(u_x - v_x)^2 + (u_y - v_y)^2}$$

Note that there is nothing special to this definition of distance: it is a direct application of the Pythagorean theorem. Besides an attracting force, Eades also introduces a repelling force $F_{rep}(u,v)$ between nonadajcent vertices u and v, defined as:

$$F_{rep}(u,v) \stackrel{\text{def}}{=} \begin{cases} 0 & \text{if } u \text{ and } v \text{ are adjacent} \\ 1/\sqrt{d(u,v)} & \text{otherwise} \end{cases}$$

With these forces defined, we now have a system of attracting and repelling vertices. When the vertices are placed randomly, it should be clear that there will generally be a lot of pushing and pulling going on. In particular, if the resulting pushing and pulling forces on a vertex are not equal, we can expect the vertex to be moving to a position in which there is more equilibrium. This behavior can be simulated by means of the following algorithm:

Algorithm 2.1 (Spring embedding):

1. Place the vertices at random locations
2. For each vertex u, calculate the resulting forces in the x and y direction, respectively:

$$F_x(u) \stackrel{\text{def}}{=} \sum_{v \neq u} \left(F_{att,x}(u,v) - F_{rep,x}(u,v) \right)$$
$$F_y(u) \stackrel{\text{def}}{=} \sum_{v \neq u} \left(F_{att,y}(u,v) - F_{rep,y}(u,v) \right)$$

3. Reposition vertex u according to:

$$u_x \leftarrow u_x + 0.1 \cdot F_x(u) \quad \text{and} \quad u_y \leftarrow u_y + 0.1 \cdot F_y(u)$$

4. Goto to Step 2. Stop after M iterations.

$F_{att,x}(u,v)$ is the attracting force *in the x direction* from neighboring vertex v, computed as:

$$F_{att,x}(u,v) \stackrel{\text{def}}{=} F_{att}(u,v) \cdot \frac{|v_x - u_x|}{d(u,v)}$$

The respective definition of $F_{att,y}(u,v)$, $F_{rep,x}(u,v)$, and $F_{rep,y}(u,v)$ is analogous.

Furthermore, we have used the notation "$x \leftarrow S$" to denote that x takes on the value resulting from evaluating expression S. So, in our example, u_x is adjusted by $0.1 \cdot F_x(u)$ units. In practice, a state very close to equilibrium is reached for at most $M = 100$ iterations.

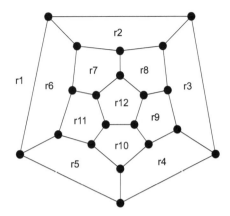

Figure 2.18: A plane graph with 12 regions.

Planar graphs

Let us now take a look at an important class of graphs where topology plays a role, namely graphs that can be drawn in such a way that no edges cross each other, so-called planar graphs.

Definition 2.15: *A **plane graph** is a specific embedding of a graph G such that no two edges intersect. If such an embedding exists, G is said to be **planar**.*

It is not difficult to see why planarity can play an important role. Consider, for example, designing a transportation network. If the corresponding graph is planar, this means that there is no need for multi-layer crossings such as bridges and tunnels. As another example, consider the design of electrical circuits, such as those for chips. In this case, it is important that the wires that connect components do not cross each other. Unfortunately, designing modern chips under the constraint that the associated graph must be planar is very difficult, if not impossible. The alternative, is to design chips as a collection of layers, each layer having an associated planar graph. We will later discuss another intriguing application of planar graphs, namely the coloring of maps. Before doing so, we first consider a number of characteristic properties of planar graphs.

When considering a plane graph, we will observe a number of regions (also called faces), which are enclosed by the edges of the graph. For example, Figure 2.18 shows a graph with 12 regions. Each region, except r_1, is enclosed by a cycle. Region r_1 is referred to as an **exterior region**; the others are **interior regions**.

A useful property of planar graphs was formulated by the famous ver-

2.4. DRAWING GRAPHS

satile Swiss mathematician Leonhard Euler (1707-1783), who is generally considered as one of the greatest mathematicians ever.

Theorem 2.7 (Euler's formula): *For a plane graph G with n vertices, m edges, and r regions, we have that* $n - m + r = 2$.

To prove this theorem, we need to consider an important property of an **acyclic graph**, that is, a graph containing no cycles, also known as a tree. Formally, we have:

Definition 2.16: *A simple, connected graph having no cycles is called a **tree**. A simple graph having only trees as its components, is called a **forest**.*

Lemma 2.1: *Any tree T with n vertices has* $|E(T)| = n - 1$ *edges.*

Proof. We prove this lemma by induction on the number of vertices. Clearly, when $n = 1$ there can be no edges and the lemma is seen to hold. Now assume the lemma holds for all trees with less than n vertices. Let H be a tree with $n \geq 2$ vertices, and edge $\langle u, v \rangle \in E(H)$. If we remove this edge, then the result will be two separate subgraphs G_1 and G_2, for otherwise $\langle u, v \rangle$ was part of a cycle. Both subgraphs are acyclic, each with less than n vertices, so that $|E(G_1)| = |V(G_1)| - 1$ and $|E(G_2)| = |V(G_2)| - 1$. Because we have not removed any vertices, we know that

$$|E(H)| = |E(G_1)| + |E(G_2)| + 1 = |V(G_1)| - 1 + |V(G_2)| - 1 + 1 = n - 1$$

which completes the proof. □

Note 2.14 (Proof techniques)
This is the first time we have encountered a **proof by induction**. This type of proof consists of two parts. First, a situation is shown to hold for some initial value n (in our example, the number of vertices for which we first consider one vertex). Then, assuming the situation is valid for $k > 1$, we prove that it also holds for $k + 1$. In doing so, we have then completed the proof.

Proof by induction is extremely important and you should make sure that you not only understand it well, but also that you are proficient in applying it. d'Angelo and West [2000] devote a complete chapter to the principle of induction and provide many examples of its use. Formally, induction is defined by considering the natural numbers, that is, $\mathbb{N} \stackrel{\text{def}}{=} \{1, 2, \ldots\}$. We then have the following important theorem.

Theorem 2.8 (Principle of induction): *Let $S(n)$ be a mathematical statement formulated in terms a natural number n. $S(n)$ is true if the following two statements are true:*

> 1. $S(1)$ is true
> 2. for any $k \in \mathbb{N}$, if $S(k)$ is true, then $S(k+1)$ is true
>
> What this theorem tells us, is that to conduct a proof by induction, we need to first show that $S(1)$ holds. Secondly, we need to show that if $S(k)$ is true, then $S(k+1)$ is also true. We can then conclude that $S(n)$ is true for any $n \in \mathbb{N}$. In practice this means that we show $S(1)$ to be true, then *assume* that $S(k)$ is true for $k > 1$, after which we need to show that $S(k+1)$ is true based on that assumption.
>
> Of course, showing that $S(k+1)$ is true is often the nasty part. A common approach is to try to reduce the situation for $k+1$ to $S(k)$. This is exactly what happened in the case of our lemma: we simply removed an edge which lead to subgraphs of smaller size for which we knew that our statement $n = m + 1$ was true. From there on, we could subsequently count the number of vertices and edges in the original graph.

Using this lemma, we can now complete our proof of Euler's formula, again by means of induction:

Proof of Theorem 2.7. The proof is by induction on r, the number of regions. If $r = 1$, then there is only a single region, which means there cannot be a region enclosed by edges of G. In other words, G must be acyclic, in which case $m = n - 1$ and thus $n - m + r = n - (n-1) + 1 = 2$. For $r = 1$ the formula is therefore seen to be true.

Now assume the formula is true for all plane graphs with less than r regions, and let G be a plane graph with $r > 1$ regions. Choose an edge e (which is not a cut edge) and consider the subgraph $G^* = G - e$. As e was part of a cycle, we will have merged two regions, reducing the total number of regions by 1. In that case, we know that Euler's formula is true, and as a consequence, $|V(G^*)| - |E(G^*)| + (r-1) = 2$. Considering that $|V(G^*)| = |V(G)|$ and $|E(G^*)| = |E(G)| - 1$, we now obtain $|V(G)| - (|E(G)| - 1) + r - 1 = |V(G)| - |E(G)| + r = 2$, completing our proof. □

Euler's formula is important as it allows us to derive a number of properties by which we can more easily determine whether a given graph is planar or not. To this end, we first prove the following:

Theorem 2.9: *For any connected simple planar graph G with $n \geq 3$ vertices and m edges, we have that $m \leq 3n - 6$*

Proof. Consider a region f in any plane graph of G. For any interior region, let $B(f)$ denote the number of edges by which f is enclosed, i.e., the length

2.4. DRAWING GRAPHS

of its "border." Obviously, $B(f) \geq 3$ for any interior region. However, with $n \geq 3$ we also have that the exterior region is "bounded" by at least 3 edges. Therefore, if there are a total of r regions, then clearly $\sum B(f) \geq 3r$. On the other hand, it is not difficult to see that $\sum B(f)$ counts every edge in G once or twice, and hence $\sum B(f) \leq 2m$, so that we obtain $3r \leq \sum B(f) \leq 2m$, and thus $r \leq \frac{2}{3}m$. From Theorem 2.7 we then derive that $m = n + r - 2 \leq n + \frac{2}{3}m - 2$, so that $m \leq 3n - 6$. □

Note that this theorem gives us a *necessary* condition for a simple graph to be planar. In other words, if we have a simple graph G for which $m > 3n - 6$, then G cannot be planar. It is not a *sufficient* condition, as we will show shortly. Furthermore, what we learn from this theorem is that a planar graph will have relatively few edges, which is intuitively clear. We can use it to prove that the complete graph on 5 vertices, that is, K_5 cannot be planar.

Corollary 2.4: *The complete graph on 5 vertices, K_5 is nonplanar.*

Proof. With $n = |V(K_5)| = 5$ and $m = |E(K_5)| = \binom{5}{2} = 10$, we have that $m \not\leq 3n - 6$, so that K_5 cannot be planar. □

Note 2.15 (More information)
There are two novelties in this proof. First, we introduced the notation $\binom{n}{k}$, which is pronounced as "n choose k," and is defined as

$$\binom{n}{k} \stackrel{\text{def}}{=} \frac{n!}{(n-k)! \cdot k!}$$

Second, we are stating that the number of edges in K_n is equal to $\binom{n}{2}$. Considering that we have n vertices in K_n, it should be clear that to construct K_n, we need to consider exactly all pairs of vertices. Obviously, there are exactly $\binom{n}{2}$ of such pairs. Another way of counting the number of edges in K_n is as follows. Assume that the vertices are labeled $\{1, 2, \ldots, n\}$. For vertex 1, we can choose from $n - 1$ vertices to join it to. After that, there are only $n - 2$ vertices to join to vertex 2 (because vertex 1 is already joined with vertex 2). For vertex 3, we can choose from $n - 3$ vertices, and so. In other words, the total number of edges in K_n is equal to:

$$|E(K_n)| = (n-1) + (n-2) + (n-3) + \cdots + 2 + 1 = \frac{1}{2}n(n-1)$$

To show that $\sum_{i=1}^{n-1} i = \frac{1}{2}n(n-1)$ is left as an exercise.

Analogous to a complete graph, we also have **complete bipartite graphs** $K_{p,q}$, which is a simple graph consisting of the two disjoint set of vertices V_1

and V_2 as in Definition 2.14 on page 46, with $p = |V_1|$ and $q = |V_2|$, and a total of $p \cdot q$ edges. An observation is now the following:

Theorem 2.10: *The complete bipartite graph $K_{3,3}$ is nonplanar.*

Proof. Because $n = |V(K_{3,3})| = 6$ and $m = |E(K_{3,3})| = 9$, we find that $m \leq 3n - 6$, so that this will not give us evidence that $K_{3,3}$ is indeed nonplanar. Instead, we need to follow a similar reasoning as for the proof of Theorem 2.9. First, note that each interior region f in any $K_{p,q}$ will necessarily be enclosed by an even number of edges. Again, if $B(f)$ denotes the number of edges enclosing interior region f, and realizing that also the exterior region will be "bounded" by at least four edges, we find that $\sum B(f) \geq 4r$, where r is the total number of regions. Because edges are counted twice, we should have that $4r \leq 2m = 18$. However, Euler's formula tells us that $r = 2 - n + m = 2 - 6 + 9 = 5$, so that $4r \not\leq 18$. Therefore, $K_{3,3}$ cannot be planar. □

> **Note 2.16** (Mathematical language)
> Indeed, as we stated above, the mere fact that $m \leq 3n - 6$ is not enough to conclude that a graph is planar. In other words, it is not a *sufficient* condition.

With these two results, we can now conclude that:

Corollary 2.5: *Any connected, simple graph having a subgraph isomorphic to either K_5 or $K_{3,3}$ cannot be planar.*

Chapter 3

Extensions

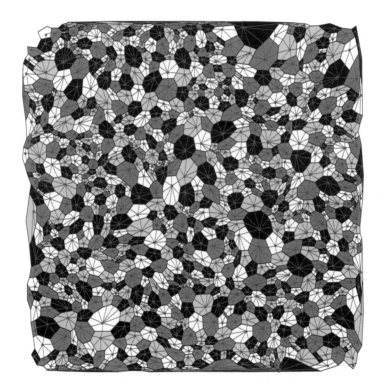

In the previous chapter we have looked only at the very basics of graphs, although it should be clear that those foundations already provide a powerful tool for modeling and analyzing real-world networks. In this chapter we consider a number of important extensions. We start with introducing graphs in which the edges are *directed*, that is, pointing from one vertex to another. Besides adding a direction to an edge, we can also associate a *weight* with an edge, which often represents some kind of cost or distance. Finally, we take a look at a specific application of graphs by which the vertices or edges are *colored*. As we shall see, colorings allow us to capture real-world situations.

3.1 Directed graphs

In the graphs we have considered so far, two vertices could be connected by one or more edges. An edge was represented by an unordered pair of vertices, such as $\langle u,v \rangle$ in the case of simple graphs. However, having no ordering is not always convenient. Consider the following examples:

- Suppose we want to model a street plan as a network. This is naturally done by representing a junction as a vertex and a street as an edge connecting two junctions. However, we need a notion of edge direction if we want to represent one-way streets.

- In social relations it is often convenient to represent the fact that Alice knows Bob, but that the opposite is not the case. In a social network this is done by representing people by vertices, and the "who knows whom" relation by means of directed edge.

- In computer networks, and notably wireless networks, links between two different nodes are often not symmetric in the sense that messages can generally be successfully sent from station A to B, but not the other way around. Modeling such a computer network is more conveniently done using directed edges.

What we are thus seeking is a way to extend graphs that we will be able to model these and similar situations.

Basics of directed graphs

The need for associating a direction with the edges of a graph leads to the notion of a directed graph, or simply digraph:

Definition 3.1: *A **directed graph** or **digraph** D consists of a collection **vertices** V, and a collection of **arcs** A, for which we write $D = (V, A)$. Each arc $a = \langle \overrightarrow{u,v} \rangle$ is* said to join vertex $u \in V$ to another (not necessarily distinct) vertex v. Vertex u is called the **tail** of a, whereas v is its **head**.

The **underlying graph** $G(D)$ of a digraph D is obtained by replacing each arc $a = \langle \overrightarrow{u,v} \rangle$ with its undirected counterpart. As we shall see in later chapters, analyzing the underlying graph is often more convenient than directly considering the original digraph. Conversely, we can transform any undirected graph G into a directed one, $D(G)$, by associating a direction with each edge. Such a digraph is also known as an **orientation**. We leave it as an exercise to prove that for a simple graph G with m edges that there are 2^m different orientations possible.

As with undirected graphs, neighbor sets play an important role in directed graphs. We make a distinction between two types of neighbors:

Definition 3.2: *Consider a directed graph D and vertex $v \in V(D)$. The **in-neighbor set** $N_{in}(v)$ of v consists of the adjacent vertices having an arc with v as its head. Likewise, the **out-neighbor set** $N_{out}(v)$ consists of the adjacent vertices having an arc with v as its tail. Formally we have:*

$$N_{in}(v) \stackrel{def}{=} \{w \in V(D) | v \neq w, \exists a = \langle \overrightarrow{w,v} \rangle : a \in A(D)\}$$
$$N_{out}(v) \stackrel{def}{=} \{w \in V(D) | v \neq w, \exists a = \langle \overrightarrow{v,w} \rangle : a \in A(D)\}$$

The set of neighbors $N(v)$ of vertex v is simply the union of its in-neighbors and out-neighbors, i.e., $N(v) \stackrel{def}{=} N_{in}(v) \cup N_{out}(v)$.

> **Note 3.1** (Mathematical language)
> Notice that the formal part of this definition is almost identical to that of the neighbor set in the case of undirected graphs. And again, it is precise, yet can seem somewhat intimidating at first sight. Informally, the in-neighbor set consists of adjacent vertices from which v can be directly reached: they are neighbors "pointing" to v. The out-neighbor set consists of vertices to which v is "pointing." These type of informal translations of mathematical definitions are important to make, and as before, you are encouraged to practice in formulating them.

A digraph is said to be **strict** if it has no loops and no two arcs with the same end points have the same orientation. Note that the notion of a strict digraph is analogous to that of a simple undirected graph. Many concepts that we defined for undirected graphs have their counterparts in digraphs. Let us start with that of vertex degree.

Definition 3.3: *For a vertex v of digraph D, the number of arcs with head v is called the **indegree** $\delta_{in}(v)$ of v. Likewise, the **outdegree** $\delta_{out}(v)$ is the number of arcs having v as their tail.*

3.1. DIRECTED GRAPHS

The concept of indegree and outdegree can sometimes play a surprisingly important role when devising or analyzing real-world networks. To give an example, suppose we are devising a communication network in which we model the case that node u can send a message directly to node v by means of an arc $a = \langle \overrightarrow{u,v} \rangle$. The indegree of node v may then indicate how many messages v can expect per time unit, also known as the rate of incoming messages. In many cases, it is desirable that this rate is limited in order to ensure that nodes are not overloaded.

In general, considering **vertex degree distributions** is an important technique for analyzing networks. A degree distribution shows how many vertices have degree $0, 1, 2, \ldots$, and so on. In many practical cases, we are often more interested in finding the distribution of the indegrees. For example, in the case of social networks, nodes with a high indegree are often considered to be important. By computing the ratio of indegrees between different nodes, we can get an impression of exactly how more important certain nodes are. We will return to vertex degree distributions extensively in later chapters.

Returning to graph-theoretical issues, it is not difficult to see that the following analogy to undirected graphs holds:

Theorem 3.1: *For any directed graph D the sum of indegrees as well as the sum of outdegrees is equal to the total number of arcs:*

$$\sum_{v \in V(D)} \delta_{in}(v) = \sum_{v \in V(D)} \delta_{out}(v) = |A(D)|.$$

Proof. Clearly, every arc in D has exactly one head and one tail. The sum of the indegrees is the same as counting all arc heads, and likewise, the sum of all outdegrees is the same as counting all tails, both being equal to the total number of arcs. □

A natural representation of directed graphs is by means of an adjacency matrix $\mathbf{A}$ in which $\mathbf{A}[i, j]$ is equal to the number of arcs joining vertex v_i to v_j. In contrast to an adjacency matrix for an undirected graph, we have the following properties in case of a directed graph:

- A digraph D is strict if and only if for all i and j, $\mathbf{A}[i,j] \leq 1$ and $\mathbf{A}[i,i] = 0$. In other words, there can be at most one arc joining any vertex v_i to another vertex v_j, and no arcs joining a vertex to itself.

- For each vertex v_i, $\sum_j \mathbf{A}[i,j] = \delta_{out}(v_i)$ and $\sum_j \mathbf{A}[j,i] = \delta_{in}(v_i)$. In other words, the sum of the entries in *row i* corresponds to the outdegree of vertex v_i, whereas the sum of the entries in *column i* equals the indegree of v_i.

Note that in contrast to undirected graphs, the adjacency matrix for a directed graph is *not* necessarily symmetric, that is, in general, $\mathbf{A}[i,j] \neq \mathbf{A}[j,i]$. Rephrasing this in natural language means that when there is an arc joining vertex v_i to v_j, then there need not necessarily also be an arc joining v_j to v_i. Taking the same graph from Figure 2.7 but now with a specific orientation, Figure 3.1 shows an example of a digraph and its adjacency matrix.

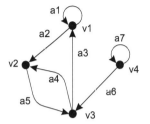

	v_1	v_2	v_3	v_4	Σ
v_1	1	1	0	0	2
v_2	0	0	1	0	1
v_3	1	1	0	0	2
v_4	0	0	1	1	2
Σ	2	2	2	1	7

Figure 3.1: A digraph with its associated adjacency matrix.

Similarly, we can represent a digraph by means of an incidence matrix $\mathbf{M}$. In this case, $\mathbf{M}[i,j]$ represents whether or not vertex v_i is incident to arc a_j. In particular:

$$\mathbf{M}[i,j] = \begin{cases} 1 & \text{if vertex } v_i \text{ is the tail of arc } a_j \\ -1 & \text{if vertex } v_i \text{ is the head of arc } a_j \\ 0 & \text{otherwise} \end{cases}$$

Unfortunately, if a digraph has loops (i.e., arcs of the form $\langle \overrightarrow{u,u} \rangle$ that join a vertex to itself), this representation will not work, as is also illustrated in Figure 3.2. Partly also for this reason, it is more common to use adjacency matrices or simply listing the arcs analogous to edge-list representations in the case of undirected graphs.

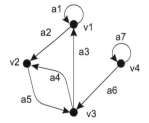

	a_1	a_2	a_3	a_4	a_5	a_6	a_7
v_1	0	1	-1	0	0	0	0
v_2	0	-1	0	-1	1	0	0
v_3	0	0	1	1	-1	-1	0
v_4	0	0	0	0	0	1	0

Figure 3.2: A digraph with its associated incidence matrix.

Connectivity for directed graphs

Connectivity is also an important concept for directed graphs. To define connectivity for digraphs, we need the equivalent notions of paths.

Definition 3.4: *Consider a digraph D. A **directed (v_0, v_k)-walk** in D is an alternating sequence $[v_0, a_0, v_1, a_1 \ldots v_{k-1}, a_{k-1}, v_k]$ of vertices and arcs from D with $a_i = \langle \overrightarrow{v_i, v_{i+1}} \rangle$. A **directed trail** is a directed walk in which all arcs are distinct; a **directed path** is a directed trail in which all vertices are also distinct. A **directed cycle** is a directed trail in which all vertices are distinct except for v_0 and v_k.*

Note that the definitions of walk, trail, path, and cycle are indeed completely analogous to those for undirected graphs. The concept of a path and a cycle are practically spoken the most important ones. We can now define the connectivity of a digraph as follows:

Definition 3.5: *A digraph D is **strongly connected** if there exists a directed path between every pair of distinct vertices from D. A digraph is **weakly connected** if its underlying graph is connected.*

It is not difficult to imagine that the concept of connectivity indeed plays an important role in directed graphs. As we explained, communication networks are conveniently modeled as directed graphs. In these networks, it is important that a message from an arbitrarily chosen node u can be routed through the network to any other node. This requirement is equivalent to stating that the associated directed graph is strongly connected. Likewise, in transportation networks it is important that for an arbitrarily chosen node we can find a route to any other node. Again, this is the same as stating that we want the associated directed graph to be strongly connected.

Note 3.2 (More information)
If being strongly connected is important you may conclude that weakly connected digraphs are not that interesting. There is one important type of weakly connected digraph: a so-called **directed acyclic graph**, or simply **DAG**. A DAG is a directed graph without any directed cycle. In practice, DAGs are also assumed to be weakly connected.

Directed acyclic graphs have many applications, of which a large number deal with expressing dependencies. For example, work plans are generally broken down into smaller units such as activities. To execute a work plan, there will be many activities that can start only after the completion of other activities. These plans are conveniently modeled as directed graphs, in which a vertex represents an activity and an arc from vertex u to v the fact that activity v can start only after u has completed. For such plans, we demand that the graph is indeed acyclic.

To test for connectivity in directed graphs, we can perform a simple reachability analysis. A vertex v in a digraph D is said to be **reachable** from vertex u, if there exists a directed (u,v)-path in D. To compute the vertices that can be reached from a given vertex u, we can proceed as follows:

Algorithm 3.1 (Reachable vertices): *Let $R_t(u)$ denote the set of reachable vertices from u found after t steps.*

1. *Set $t \leftarrow 0$ and $R_0(u) \leftarrow \{u\}$.*
2. *Construct the set $R_{t+1}(u) \leftarrow R_t(u) \cup_{v \in R_t(u)} N_{out}(v)$.*
3. *If $R_{t+1}(u) = R_t(u)$, stop: $R(u) \leftarrow R_t(u)$. Otherwise, increment t and repeat the previous step.*

This is an example of a **breadth-first algorithm**, so called because at each step *each* newly added vertex is examined. We shall discuss more of such algorithms in Chapter 4. The essence of the algorithm is simple: we systematically expand the set $R(u)$ of vertices reachable from u with any new out-neighbors that can be reached once a vertex has been added to $R(u)$. Clearly, if no new neighbors are discovered [which is when $R_{t+1}(u)$ is equal to $R_t(u)$], we will have identified all reachable vertices. Then, the digraph D will be strongly connected if and only if:

$$\forall u \in V(D) : R(u) = V(D)$$

Note that we can also apply the same method for checking the connectivity of an undirected graph. We leave the description of that algorithm as an exercise.

Note 3.3 (Algorithmics)
This algorithm is expressed rather rigorously. As before, we use the notation $x \leftarrow S$ to express that the variable x takes the value resulting from evaluating the expression S. If we were to translate this algorithm into English, we would have something like:

1. Set t to 0, and let $R_0(u)$ initially contain only u.
2. Add to $R_t(u)$ all the vertices w that can be reached by an arc from v to w, where v is already contained in $R_t(u)$. Name this new set $R_{t+1}(u)$.
3. If there are no vertices that can be added to $R_{t+1}(u)$ we're done.

Making such informal translations can considerably help in understanding an algorithm. However, it should also be clear that we need the precision of the formal notation if we are to construct a program that does the job. In fact, from

3.1. DIRECTED GRAPHS

> the formal notation we can readily derive the following fragment of **pseudo-code**. (We use N_{all} to store all out-neighbors found so far, and R_{now} for the vertices that still need to be checked.)
>
> $t \leftarrow 0; R_0(u) \leftarrow \{u\};$
> **repeat**
> $N_{all} \leftarrow \emptyset; R_{now} \leftarrow R_t(u);$
>   **while** $R_{now} \neq \emptyset$ **do**
>   **select any** $v \in R_{now}; R_{now} \leftarrow R_{now} - \{v\};$
> $N_{all} \leftarrow N_{all} \cup N_{out}(v);$
>   **end while;**
> $R_{t+1}(u) \leftarrow R_t(u) \cup N_{all}; t \leftarrow t + 1;$
> **until** $R_t(u) = R_{t-1}(u);$
>
> Pseudo-code combines concepts from programming languages with mathematical and natural-language statements. The programming-language concepts are generally used for expressing the **flow of control** in an algorithm, that is, the order in which statements need to be executed. The statements themselves are written in some convenient notation. As can be seen from this example, the next step toward an actual implementation would mostly involve programming constructs for declaring and handling sets, but is otherwise independent of the algorithm.

Instead of testing for strong connectivity, we can also ask ourselves if and how we can provide an orientation for a given (connected) undirected graph such that the resulting directed graph is strongly connected. This question is relevant, for example, when designing a traffic circulation plan in which most streets should be one-way. The following theorem gives a necessary and sufficient condition for providing such an orientation.

Theorem 3.2: *There exists an orientation $D(G)$ for a connected undirected graph G that is strongly connected if and only if $\lambda(G) \geq 2$. In other words, G cannot be 1-edge-connected.*

Proof. Let us first consider a strongly connected orientation D of G. We prove, by contradiction, that G is 2-edge-connected. To that end, assume that G is not 2-edge-connected and that the removal of $e = \langle u, v \rangle$ disconnects G, that is $G - e$ falls into two components G_1 and G_2. Clearly, we can assign only one orientation to e, that is, $D(G)$ will either contain the arc $a = \langle \overrightarrow{u,v} \rangle$ or the arc $a' = \langle \overrightarrow{v,u} \rangle$. Because all paths in G from a vertex $x \in V(G_1)$ to a vertex $y \in V(G_2)$ will contain e, it is also clear that with either orientation of e, $D(G)$ cannot be strongly connected, which violates our initial assumption. Hence, G cannot be 1-edge-connected and therefore is (at least) 2-edge-connected.

We now need to prove necessity, that is, $\lambda(G) \geq 2$, then there exists a an orientation D of G that is strongly connected. Consider a 2-edge-connected undirected graph G. From Corollary 2.3 we know that every edge of G lies on a cycle. Consider a cycle $C = [v_1, v_2, \ldots, v_n, v_1]$. We replace each edge $\langle v_i, v_{i+1}\rangle$ with an arc $\langle \overrightarrow{v_i, v_{i+1}}\rangle$ and edge $\langle v_n, v_1\rangle$ with arc $\langle \overrightarrow{v_n, v_1}\rangle$. Any edge $\langle v_i, v_j\rangle$ between nonadjacent vertices on C can be oriented arbitrarily. This situation is shown in Figure 3.3(a). Clearly, if $V(C) = V(G)$ we will have constructed a strongly connected orientation of G.

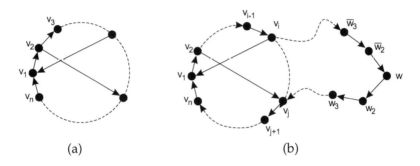

Figure 3.3: The construction of a strongly connected orientation. In (a) we have found part of the orientation by considering a cycle C. In (b), the existing orientation is extended for vertices not lying on C.

Assume $V(C) \neq V(G)$ so that we have not yet covered all vertices of G. Let w be such a vertex, i.e., $w \notin V(C)$. Because G is 2-edge-connected, we know from Corollary 2.2 that there are two edge-independent paths connecting w to v_1, as shown in Figure 3.3(b). Without loss of generality, we may assume that these two paths partly overlap with C. One path, P_1, will have the form $[w = w_1, w_2, \ldots, w_k, v_j, v_{j+1}, \ldots, v_1]$. The other will necessarily have the form $[w = \overline{w}_1, \overline{w}_2, \ldots, \overline{w}_l, v_i, v_{i-1}, \ldots, v_1]$, where $1 \leq i \leq j \leq n$. We then transform each edge $\langle w_x, w_{x+1}\rangle$ to the arc $\langle \overrightarrow{w_x, w_{x+1}}\rangle$, and each edge $\langle \overline{w}_y, \overline{w}_{y+1}\rangle$ to $\langle \overrightarrow{\overline{w}_{y+1}, \overline{w}_y}\rangle$. Again, edges between nonadjacent vertices on P_1 and P_2 may be oriented arbitrarily. It should be clear that all vertices in $W = V(C) \cup V(P_1) \cup V(P_2)$ are connected through two edge-disjoint paths in D.

If there is still a vertex in $V(G)\backslash W$, we simply repeat the procedure until all edges have been provided with an orientation. The result will be a strongly connected orientation of G. □

Again, notice how our proof consists of a part proving *sufficiency*, and a part proving *necessity*.

> **Note 3.4** (Proof techniques)
> This is typically one of those proofs where visualization is almost a necessity. In fact, the proof by itself is not even that difficult to produce once you have a fairly clear picture of what is going on. In this case, the more difficult part is providing the correct mathematical notations and statements. As we have argued before, in cases such as these it makes sense to practice reproducing the proof so that you force yourself to be precise and to get further acquainted with the language of mathematics.
>
> Another issue worthwhile noting about this proof, is that we stated that without loss of generality, we could assume that both P_1 and P_2 overlap with C. This is an important assumption: a *special* case would be when there would be no overlap. However, note that our proof also covers the cases when either one or both paths would be edge independent from C. In that case, the proposed orientation would still ensure that there is a directed path from w to v_1 and one from v_1 to w, which is exactly what we required for being strongly connected.
>
> Finally, note that we have made use of two proof techniques. To prove that G is 2-edge-connected when there is a strongly connected orientation, we applied a proof by contradiction. Proving that there is a strongly connected orientation when G is 2-edge-connected was accomplished by a proof by construction. As mentioned before, the latter has the strong advantage that we actually show *how* to obtain such an orientation.

As mentioned, digraphs play an important role when modeling real-world networks. We will come across various applications in later chapters, but notably when considering the Web in Chapter 8, it will become clear that the concepts of connectivity and (in)degree distribution play a crucial role in obtaining a deeper insight in the organization of the world's largest information system.

3.2 Weighted graphs

Let us now direct our attention to another important extension of the foundations discussed in Chapter 2, namely assigning weights to edges (or arcs). A weight is a real-valued number associated with an edge. This extension is a natural one when modeling real-world networks as graphs. For example, when modeling a railway network as a graph, railway stations are naturally represented by vertices, whereas two adjacent stations are connected by means of an edge. We then assign a weight to an edge representing the distance between those two stations.

Definition 3.6: *A **weighted graph** G is a graph for which each edge e has an associated real-valued number $w(e)$ called its **weight**. For any subgraph $H \subseteq G$, the weight of H is simply the sum of weights of its edges: $w(H) = \sum_{e \in E(H)} w(e)$.*

A commonly adopted convention for weighted graphs is to simply write that $w(\langle u,v \rangle) = \infty$ when vertices u and v are not adjacent. This also means that for each edge $e \in E(G)$ we demand that $w(e) < \infty$.

We often use weighted graphs to find subgraphs with a maximal (or minimal) weight. In particular, we can use them to determine the distance between two vertices, which is formally defined as follows.

Definition 3.7: *Consider an undirected graph G and two vertices $u, v \in V(G)$. Let P be a (u,v)-path having minimal weight among all (u,v)-paths in G. The weight of P is known as the **(geodesic) distance** $d(u,v)$ between u and v. Path P is called a **shortest path** (u,v)-path, or a **geodesic** between u and v.*

Finding shortest paths is a central problem in virtually all communication networks. Fortunately, there exists an efficient algorithm for computing the shortest paths from a given vertex u to all other vertices in a given undirected graph. Again, this is an example of a breadth-first algorithm.

The algorithm, due to the Dutch mathematician Edsger Dijkstra, was developed in 1959 and forms the core of many so-called **routing algorithms** that are used in the Internet. It is beyond doubt one of the most important algorithms in modern communication networks. The principle is as follows. Consider an undirected graph G, a vertex $u \in V(G)$, and the set $S(u)$ of vertices whose shortest path from u has already been found. In each step we, consider the set of vertices that are adjacent to some vertex in $S(u)$ but do not belong to $S(u)$ yet. We pick the one among these vertices that is closest to u and then add it to $S(u)$.

Before we formally describe the algorithm, let us consider an example. In Figure 3.4 we see a simple graph for which we want to find the shortest paths originating from vertex v_0. We start with initializing $S(v_0)$ to $\{v_0\}$ and consider the vertex that is closest to v_0. In our example, this vertex is v_3, which is subsequently added to the set $S(v_0)$. In addition, we label v_3 with (k,d), where k is the index of the vertex through which v_0 can reach v_3 (which, in this case, is v_0, i.e., $k = 0$), and d is the length of the shortest path to v_3 (with $d = 1$ in this example).

The procedure continues with identifying the vertex closest to v_0 that can be reached from any vertex in $S(v_0)$, which is now equal to $\{v_0, v_3\}$. Clearly, this is vertex v_2, which is then added to $S(v_0)$ and receiving label $(0,3)$. The next vertex to add is v_5: with $S(v_0)$ now being equal to $\{v_0, v_2, v_3\}$, the vertices reachable from $S(v_0)$ are v_1, v_4, v_5, and v_6, at distances 5 (via v_2), 6 (via v_0), 4 (via v_2), and 5 (via v_3), respectively. After adding v_5 to $S(v_0)$ and giving it label $(2,4)$, we can choose either v_1 or v_6, which are both at distance 5 from v_0. This procedure continues until all vertices from G have been added to $S(v_0)$. Let us now formally describe Dijkstra's algorithm.

3.2. WEIGHTED GRAPHS

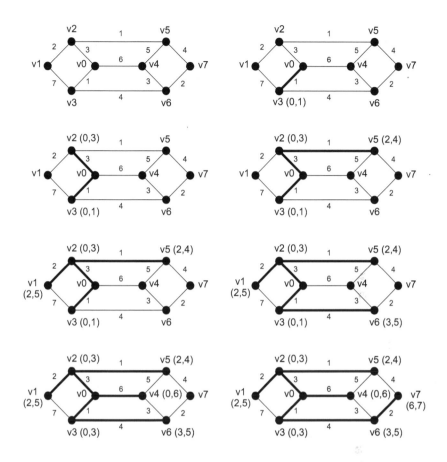

Figure 3.4: Computing the shortest paths from v_0.

Algorithm 3.2 (Dijkstra): *Consider an undirected, simple weighted graph G. Edge weights are required to be nonnegative. Consider a vertex u. We introduce the following sets and labels:*

- *Let $S_t(u)$ be the set of vertices to which a shortest path from vertex u has been found after step t.*

- *Each vertex v is assigned a label $\mathbf{L}(v) \stackrel{\text{def}}{=} (L_1(v), L_2(v))$, in which $L_1(v)$ is the vertex preceding v in the shortest (u,v)-path found so far, and $L_2(v)$ the total weight of that path.*

- *Let $R_t(u) \stackrel{\text{def}}{=} S_t(u) \cup_{v \in S_t(u)} N(v)$, with $N(v)$ denoting the neighbor set of v. In other words, $R_t(u)$ consists of all vertices in $S_t(u)$ and their neighbors.*

1. Initialize $t \leftarrow 0$ and $S_0(u) \leftarrow \{u\}$. Furthermore, for all $v \in V(G)$:

$$\mathbf{L}(v) \leftarrow \begin{cases} (u, 0) & \text{if } v = u \\ (-, \infty) & \text{otherwise} \end{cases}$$

2. For each vertex $y \in R_t(u) \setminus S_t(u)$, consider the vertices $N'(y)$ that are neighbors of y that lie in $S_t(u)$, i.e., $N'(y) \stackrel{\text{def}}{=} N(y) \cap S_t(u)$. Select $x \in N'(y)$ for which $L_2(x) + w(\langle x, y \rangle)$ is minimal. Set $\mathbf{L}(y) \leftarrow (x, L_2(x) + w(e))$.

3. Let $z \in R_t(u) \setminus S_t(u)$ for which $L_2(z)$ is minimal. Set $S_{t+1}(u) \leftarrow S_t(u) \cup \{z\}$. If $S_{t+1}(u) = V(G)$, stop. Otherwise, $t \leftarrow t + 1$, compute $R_t(u)$ again and repeat the previous step.

Note 3.5 (Algorithmics)
Admittedly, the formal description of Dijkstra's algorithm is not an easy read. This is partly caused by the fact that we need to express the flow of control, which is rather awkward. Using pseudo-code, things become much easier to read. Strictly following our previous notations, yet omitting the step counter t, we obtain the following code fragment:

$S(u) \leftarrow \{u\}$
$\mathbf{L}(u) \leftarrow (u, 0)$; for each $v \in V(G), u \neq v : \mathbf{L}(v) \leftarrow (-, \infty)$;
while $S(u) \neq V$ **do**
    $R(u) \leftarrow S(u) \cup_{v \in S(u)} N(v)$;
    for all $y \in R(u) \setminus S(u)$ **do**
        for all $x \in N(y) \cap S(u)$ **do**
            if $L_2(x) + w(\langle x, y \rangle) < L_2(y)$ **then**
                $\mathbf{L}(y) \leftarrow (x, L_2(x) + w(\langle x, y \rangle))$
            end if
        end for
    end for
    select $v \notin S(u)$ where $L_2(v)$ is minimal;
    $S(u) \leftarrow S(u) \cup \{v\}$;
end while

What this pseudo-code actually reveals is that the flow of control in Dijkstra's algorithm is actually quite intricate, yet that it can be completed separated from setting labels and such. Here's a good example where pseudo-code may help to better understand an algorithm.

Dijkstra's algorithm effectively creates what is known as a **tree** $T(u)$ that is said to be *rooted* at u, in this case meaning that only (u, v)-paths are of interest. In general, using Dijkstra's algorithm for a different vertex yields a different rooted tree. This can be readily observed when computing, for example, $T(v_4)$ from Figure 3.4, which we leave as an exercise. Note also that there may be more than one shortest path between two vertices u and v. In other words, there may be several (u,v)-paths all having the same minimal weight.

We shall return to shortest path algorithms, as well as various other tree-related problems in Chapter 5.

3.3 Colorings

As our last example of extensions to the foundations of graph theory discussed so far, we consider a simple labeling of edges and vertices known as **edge colorings** and **vertex colorings**, respectively. Colorings have interesting applications.

Edge colorings

Coloring graphs has drawn the attention from many researchers for the simple reason that there are so many applications that can be modeled using graph colorings. Coloring a graph means assigning a color to vertices or edges. In the case of edge colorings we are interested in assigning colors such that edges incident with the same vertex have different colors. Obviously, if a graph has m edges, we can use m different colors to establish a valid edge coloring. The trick is to find the minimal number of colors needed. Before discussing formalities, let's have a look at a simple, yet illustrative and realistic application discussed by Hall et al. [2001].

We consider a collection of n storage devices. For whatever reason, at a certain point it is necessary to move data between these devices. For example, after having observed the access patterns from users to data, it turns out that certain devices receive many more read/write requests than others, turning those devices into potential bottlenecks. By rearranging where data is stored, it may be possible to balance the load better and as a consequence remove bottlenecks.

This situation can be modeled as a directed graph with multiple arcs. Each storage device is represented by a vertex. We divide all data into equally sized units (which, in fact, is not unreasonable in practice, as files are generally divided into multiple blocks of data, each having the same size). If a block needs to be migrated from device i to j, we represent this by an arc $\langle \overrightarrow{i,j} \rangle$. In this way, every data block that needs to be migrated is represented by an arc. We now ask ourselves how quickly we can execute the complete rearrangement of data over the devices.

There are a few issues to consider. First, a device can be involved in only one migration at a time. In other words, if block b is being moved from i to j, then neither i nor j can be involved in migrating any other block of data. Second, we assume that all devices are connected to each other. In other words, it is possible to migrate data directly from any storage device to any other. Finally, we make the assumption that, if b is to be moved to j, then j has enough space left to store b. It is thus not necessary to first make space available on j, for example, by migrating another block b' from j to, say, device k.

To illustrate the problem at hand, consider four devices and a total of five blocks that need to be migrated as shown in Figure 3.5(a). In this case, it can be readily verified that in the final situation, there will 1 block in device 1, 1 block in device 2, 2 blocks in device 3, and 1 block in device 4. Such a migration will typically have been motivated by observing accesses for blocks, and subsequently redistributing the blocks in such a way that, for example, every device is receiving a fair number of access requests per time unit.

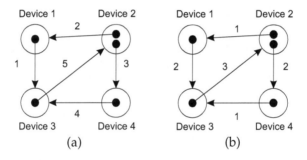

Figure 3.5: (a) A sequential migration of blocks between four devices. (b) An optimal 3-step schedule shown as an optimal edge coloring.

Obviously, we can move each block one at a time, which will take five time units. Migrations that are scheduled at time t_k are shown by the label "k" on an arc. In this case, every label as attached to an edge represents a color. The situation in Figure 3.5(a) thus reflects a situation in which we have used five different colors. Note that the requirement that a device cannot be involved in more than one migration at a time corresponds precisely to that of edge colorings: all arcs or edges incident to a vertex need to have a different color. A more efficient schedule is shown in Figure 3.5(b) in which a number of migrations take place simultaneously. The situation sketched in Figure 3.5(b) corresponds to a *minimal edge coloring* with three different colors. Specifically, we can complete the migration in three time units, instead

3.3. COLORINGS

of the original five.

Formally, edge colorings are defined as follows:

Definition 3.8: *Consider a connected, loopless graph G. G is **k-edge colorable** if there exists a partitioning of $E(G)$ into k disjoint sets $E_1, \ldots, E_k$ such that no two edges from the same E_i are incident with the same vertex.*

> **Note 3.6** (Mathematical language)
> A partitioning of a set S is formally defined as a collection of sets $S_1, \ldots, S_k$ such that
>
> - Each S_i is a subset of S, meaning that $\forall i : S_i \subseteq S$
> - These sets together constitute S, that is, $S_1 \cup S_2 \cup \cdots \cup S_k = S$, or, equivalently $\cup_{i=1}^{k} S_i = S$
> - No two sets have an element in common, which can be mathematically written as $\forall i \neq j : S_i \cap S_j = \emptyset$
>
> Now consider a graph G and a partitioning $\mathcal{E}$ of its edge set $\{E_1, \ldots, E_k\}$. Let V_i be the set of vertices formed by the end points of edges in E_i. We leave it as an exercise to show that $\mathcal{E}$ is an edge coloring of G if and only if $|V_i| = 2 \cdot |E_i|$.

As mentioned, the edge-coloring problem for graphs is finding the minimal k for which a graph G is k-edge colorable. This minimal k for a graph G is called G's **edge chromatic number**, denoted by $\chi'(G)$. If $\Delta(G)$ is the maximal degree of a vertex in graph G, it is obvious that $\chi'(G) \geq \Delta(G)$. We can even be more specific if we consider simple graphs:

Theorem 3.3 (Vizing): *For any simple graph G, either $\chi'(G) = \Delta(G)$ or $\chi'(G) = \Delta(G) + 1$.*

The proof is not difficult but somewhat involved and we omit it here. The interested reader is referred to Bondy and Murty [1976].

Vertex colorings

Perhaps more than the edge-coloring problem, researchers have paid significant attention to the vertex-coloring problem. In essence, the problem boils down to finding a coloring of the vertices of a (simple, connected) graph such that no two adjacent vertices have the same color. The problem becomes interesting when we try to use a minimal number of different colors.

Definition 3.9: *Consider a simple connected graph G. G is **k-vertex colorable** if there exists a partitioning of $V(G)$ into k disjoint sets $V_1, \ldots, V_k$ such that no two vertices from the same V_i are adjacent.*

> **Note 3.7** (Mathematical language)
> The mathematical formulation that no two vertices from the same V_i are adjacent is:
> $$\forall V_i, \forall x,y \in V_i : \nexists e \in E(G) : e = \langle x,y \rangle$$
> where $\nexists$ is to be read as "there does not exist..." In other words, for all pairs of distinct vertices in V_i there is not an edge joining those two vertices. Note that in Chapter 2 we mentioned that $\langle x,y \rangle$ is strictly speaking nothing else but stating that x and y are adjacent. If use the notation $\neg \langle x,y \rangle$ to indicate that x and y are *not* adjacent, we can simplify our mathematical formulation to:
> $$\forall V_i, \forall x,y \in V_i : \neg \langle x,y \rangle$$
> It is important that you gradually become familiar with these type of formal statements, but also that you can devise them yourself.

The vertex-coloring problem for a given graph G is finding the minimal k for which G is k-vertex colorable. This minimal k is called the **chromatic number** of G, denoted as $\chi(G)$.

Before we delve into various details, let us first consider a simple, yet illustrative application of vertex colorings: scheduling classes. We consider a set of n classes that need to be taught to a population of students. Two classes are not allowed to be scheduled during the same time slot if they are to be taught to the same group of students. The question is how to schedule the classes in the minimal number of slots.

This problem can be modeled by means of a graph G in which the n classes are represented by n vertices $v_1, \ldots, v_n$. Two vertices are connected by an edge if and only if there is a group of students to which the two classes must be taught. It is not too difficult to see that the minimal number of slots needed to teach all classes corresponds to $\chi(G)$, as we formally prove next.

Theorem 3.4: *The minimum number of time slots needed for the class-scheduling problem is the value of $\chi(G)$ of the associated graph G.*

Proof. We first prove that we need *at most* $\chi(G)$ slots to schedule all classes. From the definition of chromatic number, we know that any two vertices with the same color cannot be adjacent. This also means that the two classes associated with those two vertices need not be taken by the same group of students. Hence, they can be scheduled at the same time, that is, for the same time slot. In general, all vertices with the same color represent the set of classes that can be scheduled at the same time. This means that $\chi(G)$ slots are sufficient to schedule all classes.

3.3. COLORINGS

We now prove that we need *at least* $\chi(G)$ slots to schedule all classes. Suppose that $k < \chi(G)$ slots are sufficient. Classes in the same slot should be taught to different groups. In the graph G, this means that the vertices representing those classes should be nonadjacent. As a consequence, we should be able to use only k different colors yielding a k-vertex coloring of G, which contradicts the fact that $\chi(G)$ is minimal. □

> **Note 3.8** (Proof techniques)
> In our proof we have applied two techniques: the well-known proof by contradiction, and what is known as a **direct proof**. We have applied the latter already on several occasions, but this is the first time we mention it explicitly. As its name suggests, a direct proof is a general technique by which you show a statement to hold through straightforward deduction. In our proof, this straightforward deduction is done by simply considering the definition of the chromatic number and setting up a logical reasoning.
>
> An **indirect proof** is typically done by eliminating cases, and indeed, a proof by contradiction is an example of an indirect proof.

Vertex colorings are often used in the context of scheduling and optimization problems. Unfortunately, finding the chromatic number of a graph is, in general, a notoriously difficult problem. As with determining whether two graphs are isomorphic, we are dealing with a problem for which no known *efficient* solution exists (at least not when considering graphs for which $\chi \geq 3$). In effect, to determine the chromatic number we would, in principle, need to test all color assignments before coming to conclusions conclusions.

Fortunately, we can alleviate problems a bit: the chromatic number of a graph G is bounded by its maximal vertex degree $\Delta(G)$:

Theorem 3.5: *For any (simple, connected) graph G, $\chi(G) \leq \Delta(G) + 1$.*

Proof. We prove that the theorem holds by induction on the number n of vertices of G. For $n = 1$, we need to consider the complete graph K_1. Obviously, $\chi(K_1) = 1$ and $\Delta(K_1) = 0$, so that the theorem holds.

Now assume the theorem holds for all graphs on $k > 1$ vertices, and consider a graph G with $k+1$ vertices. Let vertex $v \in V(G)$ with $\delta(v) = \Delta(G)$. The graph $G^* = G - v$ has k vertices, so there exists a vertex coloring C^* of G^* with $\chi(G^*) \leq \Delta(G^*) + 1$ different colors. If $\Delta(G) = \Delta(G^*)$, then in the worst case, the number of colors used in G^* is $\chi(G^*) = \Delta(G^*) + 1 = \Delta(G) + 1$. Considering that v has $\Delta(G)$ neighbors, this means that there is a color available from the ones used in G^* that we can use for v and which has not been used for any of v's neighbors.

On the other hand, if $\Delta(G) > \Delta(G^*)$, then we can simply permit ourselves to introduce a new color for v and use the ones from an optimal coloring of G^* for all other vertices. At worst, we will then have that $\chi(G) = \chi(G^*) + 1 \leq \Delta(G^*) + 2$. If $\Delta(G^*) < \Delta(G)$, then the smallest value of $\Delta(G)$ for which this inequality is true, is, of course, when $\Delta(G) = \Delta(G^*) + 1$. Therefore, we know that $\Delta(G^*) + 2 \leq \Delta(G) + 1$, so that we indeed have that $\chi(G) \leq \Delta(G) + 1$. □

Coloring vertices would have perhaps been just one of those many graph-theoretical problems, if not for an intriguing conjecture that proved to be extremely difficult to tackle. Consider an arbitrary area map, such as one consisting of countries. We ask ourselves a simple question: if we are to color each country such that no two neighboring countries have the same color, how many different colors do we need at most? The answer turns out to be four, but it took more than 120 years to find it! Even worse, it took a computer program to find the answer. Many mathematicians were not amused.

Let's see what this map-coloring problem has to do with vertex colorings of graphs. First, the problem is easily translated into finding vertex colorings of a planar graph. Each country is represented by a vertex, and two vertices are joined by an edge if and only if they are neighbors (i.e., they share a border). Figure 3.6 shows the map of Europe and its corresponding planar graph.[1]

In 1852, the map-coloring problem surfaced and some specific cases were proven. However, it wasn't until 1976 that Appel and Haken [1976] actually solved it. More formally, they proved:

Theorem 3.6: *For any planar graph G, $\chi(G) \leq 4$.*

The only problem with their proof was that it was extremely difficult to verify. First, they split the problem into over 2000 different cases. Second, they wrote computer programs to test each case. This approach was received with a lot of reservations, notably also because researchers claimed that one would need to formally prove the correctness of the computer programs before considering their outcomes to be correct. It may be clear that Appel and Haken had entered the gray area between elegant mathematics and mechanical case testing by computers. So far, a "traditional" mathematical proof has not yet been found. It is worth noting that at that time it took more than 1200 hours of compute time to tackle the four-color conjecture. By now, however, there is no more debate about the correctness of the conjecture [Appel and Haken, 1986].

---
[1] For simplicity, some specific details have been omitted.

3.3. COLORINGS

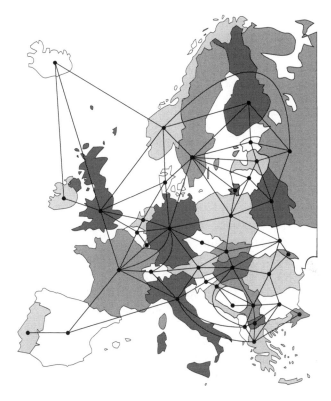

Figure 3.6: A map of Europe and its corresponding representation by a planar graph, along with a four-coloring of the vertices.

To illustrate how complications can easily sneak into mathematics, it turns out that it is relatively easy to prove that the chromatic number of a planar graph is less than or equal to 5. Before we give this proof, we need to prove the following:

Theorem 3.7: *Every planar graph G has a vertex v with $\delta(v) \leq 5$.*

Proof. For all planar graphs with $n \leq 6$ vertices, the theorem is obviously true. For planar graphs with $n > 6$, we prove the theorem by contradiction. To this end, consider a planar graph G for which $n > 6$. Let m be the number of edges of G. We know that $\sum_{v \in V(G)} \delta(v) = 2m$. Therefore, if there is no vertex with degree 5 or less, then $6n \leq 2m$. In addition, from Theorem 2.9 we know that $m \leq 3n - 6$, and thus that $6n \leq 6n - 12$. Obviously, this is false, meaning that our assumption that there is no vertex with degree 5 or less must be false as well. □

> **Note 3.9** (Proof techniques)
> Note that this proof by contradiction tells us that there must be a vertex with degree less or equal to five, but it gives us no further hints on how to find such a vertex. This is typical for **existential proofs**, in contrast to proofs by construction.

Following Chartrand [1977], we now prove the following theorem by induction on the number of vertices:

Theorem 3.8: *For any planar graph G, $\chi(G) \leq 5$.*

Proof. Let $n = |V(G)|$. For $n = 1$, the theorem is obviously true. Assume the theorem holds for all planar graphs with $k > 1$ vertices and consider a graph G with $k+1$ vertices. Let vertex v with $\delta(v) \leq 5$ (we just proved that such a vertex exists), and consider the graph $G^* = G - v$. Because $|V(G^*)| = k$, we know there exists a 5-vertex coloring of G^*, with, say, colors $c_1, \ldots, c_5$. If not all of these colors are used by the vertices in the neighbor set $N(v)$ of v, we can assign the unused color to v and will thus have constructed a 5-vertex coloring of G.

Consider the situation that all five colors have been used for coloring the vertices of $N(v)$. Note that $\delta(v) = 5$ so that we may assume that $N(v) = \{v_1, \ldots, v_5\}$ and that vertex v_i has color c_i according to a clockwise ordering of these vertices around v, as shown in Figure 3.7. We will rearrange the colors of G^* such that we can assign one of the colors c_i to v.

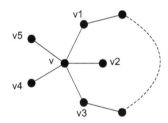

Figure 3.7: The ordering of vertices adjacent to v. Vertex v_i has color c_i.

Let us first assume that there is no (v_1, v_3)-path in G^* for which all vertices have been colored either c_1 or c_3. Now consider all paths in G^* that originate in v_1 and for which the vertices are colored either c_1 or c_3. These paths induce a subgraph H of G^*. Note that $v_3 \notin V(H)$, as this would mean that there is a (v_1, v_3)-path. For the same reason, none of v_3's neighbors can be in H, i.e., $N(v_3) \cap V(H) = \emptyset$. What we can then do is interchange the colors c_1 and c_3 in H, which leads to another 5-vertex coloring of G^*. However, in this case, vertex v_1 will be colored c_3, and none of the vertices in $N(v)$ will be colored c_1. Therefore, we can use c_1 for v.

Let us now assume that there is a (v_1, v_3)-path P in G^* for which all vertices have been colored either c_1 or c_3. Consider the cycle $[v_3, v, v_1, P]$. This cycle either encloses v_2 (as shown in Figure 3.7), or it encloses v_4 and v_5. Hence, because G is planar, there can be no (v_2, v_4)-path in G^* whose vertices are colored using only c_2 and c_4. Again, consider all paths originating in v_2 and that have either color c_2 or c_4. As before, these paths induce a subgraph H' of G^*. We interchange the colors of the vertices in H', allowing us to assign color c_2 to v, and thus leading to a 5-vertex coloring of G. □

There are many other properties related to coloring vertices, but we shall not discuss these any further. By now, it should have become clear that vertex coloring imposes a number of very difficult questions (such as efficiently finding the chromatic number of a graph), and that even under relatively favorable conditions such as planarity, taking a small step from one problem formulation ("$\chi \leq 5$") to another ("$\chi \leq 4$") can make a difference between simple and complicated solutions.

Chapter 4

Network traversal

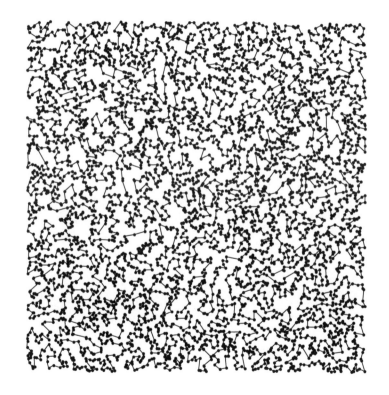

4.1. EULER TOURS

With the material presented in the previous chapters we have enough tools in our hands to start studying problems related to the traversal of networks. Network traversal problems focus on optimizing a walk that contains all vertices of a graph, also referred to as a **spanning walk**. Recall that a (v_0, v_k)-walk was defined as an alternating sequence

$$W = [v_0, e_1, v_1, e_2 \ldots v_{k-1}, e_k, v_k]$$

of vertices and edges, where edge $e_i = \langle v_{i-1}, v_i \rangle$.

One category of spanning walks that we'll consider is the one containing closed walks that also traverse each edge in a graph. These walks are also known as **tours**. An important question is to find tours in which edges are additionally crossed as few times as possible. Another important category is formed by spanning cycles. In other words, closed spanning walks in which all vertices are distinct. This so-called **Hamilton cycles** play a crucial role when we also try to minimize the total distance covered, which occurs when considering weighted graphs. Let us take a closer look at these two types of spanning walks.

4.1 Euler tours

We start our discussion with probably one of the oldest graph-theoretical problems: is it possible to traverse a graph such that all the edges are crossed exactly once? Of course, this was not how the original problem was formulated. The problem originated in the city of Königsberg (now called Kaliningrad) that was divided by the river Pregel. The several parts of the city were connected by means of seven bridges, as shown in Figure 4.1. The population of Königsberg had been amusing themselves for some time with a simple question: is it possible to walk through the city and cross each of the bridges exactly once? The answer is simply "no," but in order to understand why, we need graph theory.

Of course, if we were dealing with a puzzle applicable only to the old city of Königsberg, one could justifiable question whether it should deserve any serious attention at all. However, it turns out that the problem is easily generalized to other situations. An important one that we will discuss below is finding a spanning walk that covers every street of a city, but such that each street is preferably passed through at most once. This is the same as finding a tour with minimal total weight, where weight is defined by the length of a street. As said, we return to this important application below, after discussing some basic issues.

Figure 4.1: The seven bridges crossing the river Pregel in Königsberg.

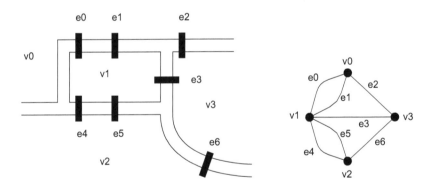

Figure 4.2: The bridges of Königsberg modeled as a graph.

Constructing an Euler tour

Returning to the seven bridges of Königsberg, we can model the problem by representing each area separated by a bridge as a vertex, and each bridge by an edge connecting two separated areas, leading to the graph (with multiple edges) shown in Figure 4.2. The people of Königsberg were interested in finding a specific **tour**:

4.1. EULER TOURS

Definition 4.1: *A **tour** of a graph G is a (u,v)-walk in which $u = v$ (i.e., it is a closed walk) and that traverses each edge in G. An **Euler tour** is a tour in which all edges are traversed exactly once.*

Euler tours were named after the Swiss mathematician Leonhard Euler who initially solved the problem of the Königsberg bridges. To this end, he proved the following theorem:

Theorem 4.1: *A connected graph G (with more than one vertex) has an Euler tour if and only if it has no vertices of odd degree.*

Proof. First, assume that P is an Euler tour of G, originating and ending in, say, vertex v. Consider a vertex u different from v. Obviously, u lies on P and for each edge $\langle w_1, u \rangle \in E(P)$ that is used for "entering" u, there is a unique other edge $\langle u, w_2 \rangle$ traversed for "leaving" u. Moreover, because these edges are traversed exactly once, edges for entering u are always uniquely paired with edges for leaving u. Hence, the degree of u must be even. By a similar reasoning, the degree of v must also be even. We conclude that all vertices of G have even degree.

Conversely, assume that all vertices of G are of even degree. We now need to prove that G has an Euler tour. To this end, select an arbitrary vertex v and construct a trail P by subsequently traversing edges until it is no longer possible to traverse an edge not belonging to P. Let w be the vertex where P ends. If $w \neq v$, then clearly we have "entered" w once more than we have "left" it, meaning that $\delta(w)$ is odd. This violates our assumption, hence $w = v$ and hence P must be a *closed* trail.

If $E(P) = E(G)$ we have just constructed an Euler tour and we're done. Now assume $E(P) \neq E(G)$, that is $E(P) \subset E(G)$. Because G is connected, there is a vertex u of P incident with edges that are not part of P. Consider the induced subgraph constructed by simply removing all edges that are part of P: $H \stackrel{\text{def}}{=} G - E(P)$. Note that H may be disconnected. Because every vertex in G has even degree, but also every vertex in P, so will every vertex in H have even degree. Let component H' contain u. Again, construct a (closed) trail P' in H' originating in u until no more edges can be added that are not yet contained in P'. Because $|E(P')| > 0$, merging P and P' will yield a *larger* trail in G. If this larger trail does not contain all edges of G, we repeat the procedure until we have constructed a closed trail containing all edges of G. This trail will form an Euler tour. □

Note 4.1 (Proof techniques)
Our proof by construction uses an important proof technique, called **extremal-**

> **ity** [West, 2001]. The essence of this technique is that we consider extreme cases, such as a path or trail of maximal length. Note that in our example, the mere fact that we construct P such that it is indeed of maximal length leads us to conclude that it is a *closed* trail. There are many other situations in which exploring extremality is necessary to draw conclusions and we will encounter more examples throughout the text.

Defining an **Euler trail** as a (u,v)-trail of a connected graph G that traverses all edges exactly once, it is not difficult to see that the following statement is true:

Theorem 4.2: *A connected graph G (with more than one vertex) has an Euler trail if and only if it has exactly two vertices of odd degree. Moreover, the trail originates and ends in the vertices of odd degree.*

Proof. First, let P be an Euler trail originating in u and ending in v. By the same reasoning as in the previous proof, all vertices except u and v must be of even degree.

Conversely, assume G has exactly two vertices u and v of odd degree. Consider the graph G^* constructed from G by adding an edge $e = \langle u,v \rangle$. All vertices in G^* will now have even degree. Because G^* is obviously also connected, we know that G^* has an Euler tour P. Removing e from P yields an Euler trail for G. $\square$

So far, we have provided only some necessary and sufficient conditions for a graph to be Eulerian. What is missing, of course, is a procedure by which we can construct an Euler tour (if one exists). The most widely known algorithm that accomplishes such a tour is due to a French mathematician, Fleury.

Algorithm 4.1 (Fleury): *Consider an Eulerian graph G.*

1. *Choose an arbitrary vertex $v_0 \in V(G)$ and set $W_0 = v_0$.*
2. *Assume that we have constructed a trail*

$$W_k = [v_0, e_1, v_1, e_2, v_2, \ldots, v_{k-1}, e_k, v_k].$$

Choose an edge incident to v_k, but which is not yet part of W_k, that is, $e_{k+1} = \langle v_k, v_{k+1} \rangle$ and $e_{k+1} \in E(G) \backslash E(W_k)$. In addition, make sure that e_{k+1} is not a cut edge of the induced subgraph $G_k = G - E(W_k)$, unless there is no other option.

3. We now have a trail W_{k+1}. If there is no edge $e_{k+2} = \langle v_{k+1}, v_{k+2}\rangle$ to select from $E(G)\backslash E(W_{k+1})$, stop. Otherwise, repeat the previous step.

Obviously, Fleury's algorithm constructs a trail in G: at no point will an edge be selected that is already part of the walk W_k constructed so far. Hence, W_k must be a trail. That the algorithm actually constructs an Euler tour is formalized in the following theorem (see also Bondy and Murty [1976]).

Theorem 4.3: *A trail constructed by Fleury's algorithm in an Eulerian graph G is an Euler tour of G.*

Note 4.2 (Algorithmics)
Before we delve into the details of this theorem, note that there is something special about it: it states that Fleury's algorithm is correct. As a consequence, if we prove this theorem, we will have shown that Fleury's algorithm indeed finds an Euler tour if one exists. Such theorem/proof combinations form a fundamental component of algorithm design in computer science. However, it is important to make a distinction between the correctness of an algorithm and the correctness of a program that *implements* that algorithm. In the latter case, we need to take into account the fact that a program is executed by a computer and that the statements we are using having precise meaning, that is, have formal semantics.

Proof (*). Let's first consider a trail W_n constructed by means of Fleury's algorithm that contains all edges of G. Assume that this trail starts in v_0 and ends in v_n. We need to show that W_n is a *closed* trail, i.e., that $v_0 = v_n$. To this end, consider the induced subgraph $G_n = G - E(W_n)$. Because W_n consists of all edges in G, each vertex in G_n must have degree 0. In particular, this is true for vertices v_0 and v_n. If $v_0 \neq v_n$, then they can only have odd degrees in G, which is impossible, because we know that G is Eulerian and thus that all vertices have even degree. Therefore, W_n must be a closed trail and thus an Euler tour.

Now suppose that W_n is not an Euler tour of G. Again, let W_n be equal to the sequence $[v_0, e_1, v_1 \ldots v_{n-1}, e_n, v_n]$. Not being an Euler tour means that we were no longer able to select any edges incident with v_n that had not already been selected. A few observations are important.

- We necessarily have that $v_0 = v_n$, for if this were not the case and there were no more edges incident with v_n to select, then following the same reasoning as before, $\delta(v_n)$ would be odd, and thus G would not be Eulerian.

- Let E_n be the edges that are not part of W_n, i.e., $E_n \stackrel{\text{def}}{=} E(G)\backslash E(W_n)$. Because W_n is assumed not to be an Euler tour, we must have that $E_n \neq \emptyset$. Let S be the set of vertices incident with edges from E_n. Some of these vertices belong to W_n, and others do not. Note that $v_n \notin S$, for otherwise this would mean that it would be incident to an edge that is not in W_n, meaning that W_n could have been expanded.

- Let $\overline{S} = V(G)\backslash S$. Note that vertices in $\overline{S}$ are not incident with edges in E_n, and thus are incident *only* with edges from W_n. In particular, $v_n \in \overline{S}$.

- Because all vertices in W_n have even degree, so will all the vertices in the induced graph $G_n \stackrel{\text{def}}{=} G[F_n]$.

- Consider a vertex u from $G_n[S]$. By definition, u is incident with an edge from E_n. Because G is Eulerian, the degree $\delta_G(u)$ of u in G is even. Also, we just observed that $\delta_{G_n}(u)$ is even. This can only mean that the degree $\delta_{G_n[S]}(u)$ of u in in the induced subgraph $G_n[S]$ of G_n is even as well.

Let m be the largest index such that $v_m \in S$ and $v_{m+1} \in \overline{S}$. In other words, v_m is the "last" vertex of W_n that is still in S, and thus incident with an edge that is not part of W_n. All other vertices $v_{m+1}, \ldots, v_n$ are in $\overline{S}$ and thus incident *only* with edges of W_n.

Now consider edge $e_{m+1} = \langle v_m, v_{m+1} \rangle$. This edge is the only edge in G_m between vertices in S and $\overline{S}$. To see this, assume there is another such edge e' in W_m. Note that because e' is incident with a vertex from S, $e' \notin E(W_n)$. On the other hand, if one of its end points belongs to $\overline{S}$, then e' would necessarily belong to $E(W_n)$, which by construction is impossible. In other words, both the end points of e' must belong to S, and hence, no e' exists. This also means that e_{m+1} is a cut edge of G_m.

Let e be any other edge in G_m incident with v_m. In Fleury's algorithm we prefer the selection of edges that are not cut edges. Because we selected e_{m+1}, which is a cut edge, e must also be a cut edge of G_m. It is then surely also a cut edge of the induced subgraph $G_m[S]$. Because $G_n \subset G_m$, we also have that $G_m[S] = G_n[S]$. As noted, all vertices in $G_n[S]$ and thus also in $G_m[S]$ have even degree. However, in a graph with only even-degree vertices, there cannot be a cut edge (which we leave as an exercise to the reader).

We have now established a contradiction based on the assumption that W_n is not an Euler tour of G. In other words, our assumption can only be false, which completes the proof. □

4.1. EULER TOURS

> **Note 4.3** (Study tip)
> Obviously, this is not an easy proof. However, despite its complexity, it is important to understand and be able to reproduce it, for it will force you to consider every detail when making a next step. At the same time it is important to grasp the big picture, namely that the construction of the proof is toward reaching a contradiction based on the fact that Fleury's algorithm prescribes that we should preferably not select a cut edge. By showing that there was no other choice (i.e., e_{m+1} is necessarily a cut edge), yet at the same time there must have been an alternative edge that was not a cut edge, we arrive at a contradiction. This contradiction tells us that when executing Fleury's algorithm, we are constructing an Euler tour, if one exists.

To see how Fleury's algorithm works, consider the graph in Figure 4.3. At each step, the bold-faced edge is added to the trail W_k. When cut edges *incident with* v_k appear in G_k, they are marked as a dashed line. These are the ones that we should prefer not to choose, but sometimes there is just no alternative. Although Fleury's algorithm is apparently elegant and simple, the difficulty in its practical execution is determining whether a selected next edge is a cut edge or not. It is for this reason that more efficient algorithms have been developed.

The Chinese postman problem

Let us now consider a practical application of Euler's research: the **Chinese postman problem**, so-called because it was first postulated by the Chinese mathematician Kuan [1962]. This problem is more general and also more complicated than that of finding an Euler tour. Consider a weighted graph G in which each edge has a nonnegative weight. The problem is to find a closed walk $W = [v_0, e_1, v_1 \ldots v_{n-1}, e_n, v_n]$ that covers all edges of G, but with minimal weight. In other words, $E(W) = E(G)$ and $\sum_{i=1}^{n} w(e_i)$ is minimal. Note that we do not demand that each edge is traversed exactly once, for in that case we would have an Euler tour, and obviously, such a walk would automatically have minimal weight. Instead, we are aiming for a closed walk such that if it is necessary to cross an edge more than once, that the walk is such that the total weight is kept as low as possible.

The Chinese postman problem is a generalization of many traversal problems. Consider the following examples.

Routing garbage trucks: In order to collect the garbage in a specific neighborhood, garbage cans are placed on the curb once a week to be emptied by trucks. An optimal route for a truck consists of passing through

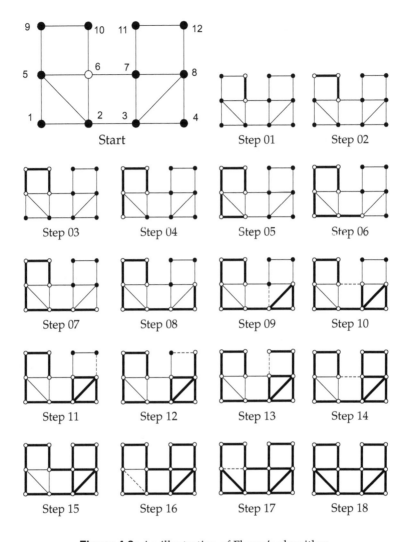

Figure 4.3: An illustration of Fleury's algorithm.

each street at least once, and possibly more, but in such a way that the total elapsed distance is minimal.

In this example, we model the neighborhood as an undirected graph in which each junction is represented by a vertex and a street as an edge with its weight corresponding to the length of the street. A variation of the problem is to allow a truck to start and end at a different location. In that case, the walk need not be closed, yet we still need to make sure that every edge is crossed at least once.

Routing a postman: Somewhat similar is determining an optimal route for a postman. However, in this case we need to take into account that streets normally have houses on both sides of a road. Rather than letting a postman cross the street from one side to the other all the time, we assume that he first delivers the mail to one side, and later to the other.

In this case, a junction is again represented by a vertex, yet a street with houses on both sides is represented by two edges, each edge effectively representing one row of houses.

Checking a Web site: Typically, a Web site consists of numerous pages, in turn containing links to each other. As is so often the case, most Web sites are notoriously poor at having their links maintained to the correct pages. This is often due to the simple reason that so many people are responsible for maintaining their part of a site. Apart from links that are broken (i.e., refer to nonexisting pages), it is often necessary to *manually* check how pages are linked to each other.

Graph theory can help by modeling a Web site as an undirected graph where a page is represented by a vertex and a link by an edge having weight 1. Note that we are not using a directed graph, as we may need to cross a link in reverse order, for example, when going back to the original page. If a site is to be manually inspected, then we are seeking a solution to navigate through a site, but with preferably crossing a link at most once. This is now the same as finding a directed walk containing all edges of minimal length.

Other examples easily come to mind, and some less obvious ones are described by Thimbleby [2003] (which includes the case of navigating through a Web site). These examples should make clear that we may sometimes need to traverse an edge twice. Formally, these means that for a closed walk $W = [v_0, e_1, v_1 \ldots v_{n-1}, e_n, v_n]$ to be minimal, it may occur that for some $i \neq j, e_i = e_j$.

In order to solve the Chinese postman problem, we proceed by transforming a non-Eulerian graph into a Eulerian one by simply *duplicating* edges. Duplicating an edge $e = \langle u, v \rangle$ means that we simply add an extra edge $e^* = \langle u, v \rangle$ with the same weight as e. The trick, of course, is to duplicate as few edges as possible and such that the added total weight of the resulting graph is minimal. Once we have transformed the original graph into a Eulerian one, we can apply Fleury's algorithm to find an Euler tour. Note that by ensuring that the total weight of the transformed graph is minimal, we also ensure that our Euler tour in the transformed graph is minimal.

Unfortunately, transforming a graph to a Eulerian one that has as less weight as possible is not trivial. For example, suppose that edge $e = \langle u, v \rangle$ is incident with a vertex v with odd degree and vertex w with even degree. Duplicating e will force us to subsequently reconsider vertex w, which in the new situation will then have odd degree. A general solution, but which is too complicated for our purposes to describe here, is given by Edmonds and Johnson [1973]. A special case that is easy to solve is when there are only two vertices having odd degree, say u and v. We can then use Dijkstra's algorithm to find a (u, v)-path having minimal weight, and subsequently duplicate each edge on that path. We leave it as an exercise to show that the result is indeed a minimum-weight Eulerian graph.

This approach can be easily generalized. Recall from Chapter 2 that every graph has an even number of vertices with odd degree, say $2k$. What we are seeking are k paths each connecting two odd-degree vertices such that no two paths have a source and destination vertex in common, and such that the sum of their respective weights is minimal. Following Gibbons [1985], we tackle this problem as follows.

Algorithm 4.2 (Chinese postman): *Consider a weighted, connected graph G with odd-degree vertices $V_{odd} = \{v_1, \ldots, v_{2k}\}$ where $k \geq 1$.*

1. *For each pair of distinct odd-degree vertices v_i and v_j, find a minimum-weight (v_i, v_j)-path $P_{i,j}$.*

2. *Construct a weighted complete graph on $2k$ vertices in which vertex v_i and v_j are joined by an edge having weight $w(P_{i,j})$.*

3. *Find the set E of k edges $e_1, \ldots, e_k$ such that $\sum w(e_i)$ is minimal and no two edges are incident with the same vertex.*

4. *For each edge $e \in E$, with $e = \langle v_i, v_j \rangle$, duplicate the edges of $P_{i,j}$ in graph G.*

The resulting graph G^* is Eulerian with minimal weight, for which we then apply Fleury's algorithm to find a minimum-weight Euler tour.

Let's consider a simple example from Gibbons [1985] to demonstrate this algorithm. Figure 4.4(a) shows our initial graph with odd-degree vertices $v_1, v_2, v_3,$ and v_4. We first find minimum-weight paths between all these vertices. It is not difficult to verify that the following paths indeed have minimal weight:

$P_{1,2} = [v_1, v_2]$ (weight: 3) $\qquad P_{2,3} = [v_2, u_3, u_5, u_4, v_3]$ (weight: 5)
$P_{1,3} = [v_1, u_2, v_3]$ (weight: 3) $\qquad P_{2,4} = [v_2, u_6, v_4]$ (weight: 2)
$P_{1,4} = [v_1, u_1, u_5, v_4]$ (weight: 5) $\qquad P_{3,4} = [v_3, u_4, u_5, v_4]$ (weight: 4)

4.1. EULER TOURS

Our next step is consider the weighted complete graph on the four vertices v_1, v_2, v_3, and v_4 as shown in Figure 4.4(b). We are seeking to find a set of two edges such that their total weight is minimal, and such that they do no have any end points in common. This is achieved by the set $\{\langle v_1, v_3\rangle, \langle v_2, v_4\rangle\}$, corresponding to the two paths $P_{1,3}$ and $P_{2,4}$. The edges of these two paths are then duplicated, leading to the Euler graph with minimal weight as shown in Figure 4.4(c).

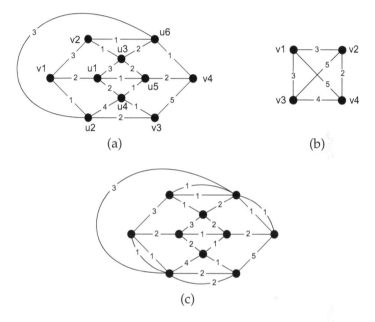

Figure 4.4: An example of solving the Chinese postman problem. (a) The initial graph, (b) finding the optimal paths, (c) the expanded graph.

Note 4.4 (More information)
The solution to the Chinese postman problem builds on an important topic in general graph theory, namely that of **matchings**. A matching M in a graph G is a subset of the edges of G such that no two edges from M are incident with the same vertex. Matchings are typically applied to situations in which we need to team up pairs of some sort, and where each pair is subject to a constraint.

Consider, for example, a group of n people $p_1, \ldots, p_n$ and m tasks $t_1, \ldots, t_m$, with $n \geq m$. A person p_i can fulfill task t_j with a certain expertise $e_{i,j} \in [0,1]$, where the value 0 represents the case that p_i cannot fulfill t_j. Assume that for each task there is at least one person who can fulfill that task. We ask ourselves what the best assignment of people to tasks is. This situation can be

> modeled by means of a weighted bipartite graph, for which we are then seeking a maximum-weight matching.
>
> In the case of the Chinese postman problem, we are actually looking for a **perfect matching**: a matching M such that every vertex in G is incident with an edge from M. There are various solutions to finding optimal (weighted) matchings, but we will not go into further details here. The interested reader is referred to Gibbons [1985].

4.2 Hamilton cycles

Where Euler tours focus on traversing every edge in a graph, **Hamilton walks** deal with traversing every vertex in a graph. In this section we concentrate on the problem of trying to construct a (closed) walk such that every vertex is visited exactly once. As we shall see, not only is this an important problem, it also turns out to be notoriously difficult if we want to optimize on the distance traveled.

Properties of Hamiltonian graphs

We start with precisely defining what a Hamiltonian graph is, along with a number of example applications.

Definition 4.2: *Consider a connected graph G. A **Hamilton path** of G is a path that contains every vertex of G. Likewise, a **Hamilton cycle** is a cycle containing every vertex of G. G is called **Hamiltonian** if it has a Hamilton cycle.*

What makes the issue of (non)Hamiltonian graphs so difficult is that, in contrast to Euler tours, there is no known efficient procedure by which one can in general determine whether a graph is Hamiltonian or not. On the other hand, finding Hamilton cycles, or closed trails that minimize the number of duplicate visits to a vertex is important. To illustrate, consider the following two problems, which are representative for a wide range of applications.

Transportation problems: Consider scheduling a service that needs to pick up people at n different locations. The problem is to find the most efficient route (e.g., expressed in the smallest traveling distance) such that all n locations are visited. This problem can be formulated in terms of a road map with locations represented as vertices and roads between pairs of locations as weighted edges. We are interested in finding a minimal weighted Hamiltonian subgraph containing all vertices, possibly after expanding the graph to account for traversing an edge more

4.2. HAMILTON CYCLES

than once. There are many variations on such transportation problems, of which a nice overview is provided by Applegate et al. [2007]. We return to this problem later in this chapter.

Drilling holes: There are many cases in which we need to drill holes in a board, such as for electrical circuits. This requires the scheduling of a drilling machine by which holes are drilled one by one. To minimize the time it takes to drill all holes, we should minimize the distance that the machine (or equivalently, the board) needs to make when moving from hole to hole. We can model this problem as a complete graph with the vertices forming the holes to be drilled and the weight on each edge representing the geometric distance of the edges two ends on the board. An optimal schedule corresponds to a minimal weighted Hamilton cycle. To illustrate, Figure 4.5(a) shows an example in which some 2400 points need to be drilled into a board. Figure 4.5(b) shows one possible schedule, whereas Figure 4.5(c) shows an optimal solution in which the machine needs to "travel" half the distance of the previous schedule. The example is discussed in more detail by Grötschel and Padberg [1993].

These two examples are instances of what is known as the **traveling salesman problem**. As mentioned, a serious issue is that there are no known efficient solutions for determining whether a graph is Hamiltonian or not. Worse, if we are interested in finding a minimal-weighted Hamilton cycle, we will have a tough problem to solve as it will most likely require a lot of computational resources. Considering the many applications related to traveling salesman problems, it should come as no surprise that researchers and practitioners have paid considerable efforts to finding efficient methods for (near-)optimal solutions.

Fortunately, there are some reasonable starting points to finding good solutions. For one, we have the following necessary condition for a graph to be Hamiltonian: if we consider a subset S of the vertices of a graph and subsequently remove those vertices, the graph should fall apart into *at most* $|S|$ components. More formally:

Theorem 4.4: *If graph G is Hamiltonian, then for every proper nonempty subset $S \subset V(G)$, we have that $\omega(G - S) \leq |S|$.*

Proof. Let C be a Hamilton cycle of G. If we consider any proper nonempty subset $S \subset V(G)$, then obviously, because every vertex is visited exactly once, the number of components in $C - S$ will be less or equal to $|S|$. However, because C contains all vertices of G, we also have that $\omega(G - S) \leq \omega(C - S)$, which completes the proof. □

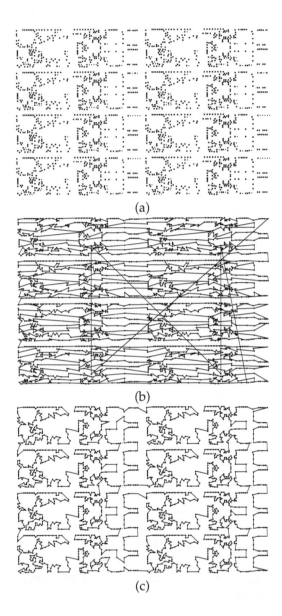

Figure 4.5: An example of scheduling a drilling machine with (a) the holes that need to be drilled, (b) a schedule, and (c) an optimal schedule. Taken with permission from [Grötschel and Padberg, 1993].

4.2. HAMILTON CYCLES

Note 4.5 (More information)
This is one of those examples where a simple diagram helps to understand what is going on. Figure 4.6 shows a graph G and an arbitrary set S of vertices from G. We have also sketched a Hamilton cycle C, which runs through every vertex in S. Effectively, we split the cycle C into alternating segments $S_1, \bar{S}_1, S_2, \bar{S}_2, \ldots, S_n, \bar{S}_n$, each segment S_i consisting of a number of consecutive vertices from S, and each segment $\bar{S}_i$ consisting of consecutive vertices *not* in S. In the "worst" situation, each subgraph induced by a segment $\bar{S}_i$ is a component of the graph $G - S$, i.e., $G[\bar{S}_i]$ is disconnected from the other parts of $G - S$. The maximal number of segments consisting of vertices outside S that we can obtain, is when each segment S_i consists of exactly 1 vertex. Hence, this maximal is equal to $|S|$.

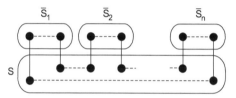

Figure 4.6: Segmentation of a Hamilton cycle for an arbitrary set S of vertices.

The previous theorem provides us with a necessary condition for a graph to be Hamiltonian. In 1952, the mathematician Gabriel Dirac proved the following sufficient condition, which essentially states that a graph is Hamiltonian if each vertex is connected to at least half of the other vertices

Theorem 4.5 (Dirac): *If G is a simple graph with $n = |V(G)|$ vertices, $n \geq 3$ and each vertex v has degree $\delta(v) \geq n/2$, then G is Hamiltonian.*

Proof. A relatively simple proof is by contradiction: assume the theorem is false. Let G be a non-Hamiltonian graph with $n \geq 3$ vertices and for which $\delta(v) \geq n/2$ for each of its vertices. Moreover, assume that G has a maximal number of edges, i.e., adding a single edge (while keeping G simple) would make it Hamiltonian. Let u and w be two nonadjacent vertices. By construction of G we know that if we add an edge $e = \langle u, w \rangle$, the resulting graph G^* would be Hamiltonian, and thus there exists a Hamilton path (u, w)-path P in G with $[u = v_1, v_2, \ldots, v_n = w]$, as shown in Figure 4.7(a).

Now consider the following two sets of vertices:

$$S = N(u) = \{v_i | \langle u, v_i \rangle \in E(G)\} \text{ and } T = \{v_i | \langle v_{i-1}, w \rangle \in E(G)\}$$

S consists of the neighbors of u, whereas T consists of the successors on P of neighbors of w. Note that $|S| \geq n/2$. Likewise, because P contains

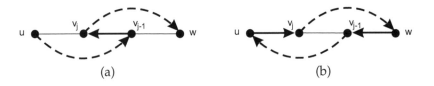

Figure 4.7: (a) A Hamilton path in G, and (b) the constructed Hamilton cycle in G.

all vertices in G, T contains as many elements as there are edges $\langle v_{i-1}, w \rangle$, which corresponds to $\delta(w)$. This means that $|T| \geq n/2$. Furthermore, vertex u is not contained in S (because it cannot be a neighbor of itself), nor is it contained in T (which contains only *successors* of other vertices on P). In other words, $S, T \subseteq \{v_2, \ldots, v_n\}$, which, together with the fact that $|S| + |T| \geq n$, means that the two sets have at least one vertex in common. Let this be vertex v_j. We now have the situation that v_j is a neighbor of u, and that v_j's predecessor v_{j-1} is a neighbor of w. But in that case, we can construct the Hamilton cycle $[u = v_1, v_j, v_{j+1} \ldots v_n = w, v_{j-1}, v_{j-2} \ldots v_1 = u]$, shown in Figure 4.7(b). Note that this cycle does *not* contain edge $\langle u, w \rangle$. In other words, we have just shown that G is Hamiltonian, which contradicts our initial assumption. This means that there is no vertex $v_j \in S \cap T$ and thus $|S \cap T| = 0$. On the other hand, we know that $u \notin S \cup T$, so that $|S \cup T| < n$. This now brings us to:

$$\delta(u) + \delta(w) = |S| + |T| = |S \cup T| + |S \cap T| < n$$

which cannot be true, meaning that we cannot assume the theorem is false. □

Note 4.6 (Proof techniques)
It is interesting to note that the proof of Dirac's theorem merely states that a Hamilton cycle exists. It does not explain how to construct such a cycle. Again, we see an important concept in mathematical proof techniques: the distinction between an existential proof and a proof by construction. There are many cases in which we know that a solution to a (graph theoretical) problem exists, but that fruitless attempts have been made to find a specific solution. However, for Dirac's theorem, a (nontrivial) proof by construction does indeed exist. We refer the interested reader to Dharwadker [2004]. Furthermore, note that we have again made use of extremality in our proof, in this example by assuming a maximal graph that was not Hamiltonian.

4.2. HAMILTON CYCLES

> **Note 4.7** (More information)
> Dirac's theorem provides a sufficient condition for a graph to be Hamiltonian. Several attempts have been made to provide a weaker sufficient condition, that is, a condition that can be met by a larger number of graphs. When looking carefully at the proof, we see that we never actually use the requirement that $\delta(v) \geq n/2$, but rather that $\delta(u) + \delta(v) \geq n$. Furthermore, strictly speaking we never used the requirement that G needed to be a maximal non-Hamiltonian graph. Instead, we only needed the property that the graph $G + \langle u, v \rangle$ was Hamiltonian. This leads to the following theorem:
>
> **Theorem 4.6** (Ore): *Let G be a simple graph with n vertices. If u and v are distinct, nonadjacent vertices with $\delta(u) + \delta(v) \geq n$, then G is Hamiltonian if and only if $G + \langle u, v \rangle$ is Hamiltonian.*
>
> As you may imagine, the proof is very much like that of Dirac's theorem. Using this theorem, another sufficient condition was formulated, based on what is known as the closure of a graph:
>
> **Definition 4.3:** *Consider a graph G with n vertices. The **closure** of G is obtained by iteratively joining each nonadjacent pair of vertices u and v for which $\delta(u) + \delta(v) \geq n$, until no such pairs exist anymore.*
>
> We can then simply prove the following theorem by applying Ore's theorem every time we add an edge in the construction of the closure of a graph:
>
> **Theorem 4.7** (Bondy-Chvátal): *A simple graph G with n vertices is Hamiltonian if and only if its closure is Hamiltonian.*

Finding a Hamilton cycle

Let us now concentrate some more on actually finding Hamilton cycles (in a simple graph). As we've mentioned before, determining whether a graph is Hamiltonian is a notoriously difficult problem in the sense that there is no known computationally efficient algorithm. In essence, this means that we can follow only a trial-and-error approach when attempting to find a Hamilton cycle, or simply doing it brute force by trying to find *all* cycles. To illustrate the latter, we can try to systematically find all cycles by means of an algorithm akin to the one for determining reachable vertices in a directed graph.

We start with randomly selecting a vertex, say v_1, and construct the set of reachable vertices as $R([v_1]) = N(v_1)$, where $[v_1]$ stands for the sequence consisting only of vertex v_1. For each vertex $u \in R([v_1])$ we then construct the set $R([v_1, u]) = N(u) \backslash \{v_1\}$. In other words, $R([v_1, u])$ consists of all

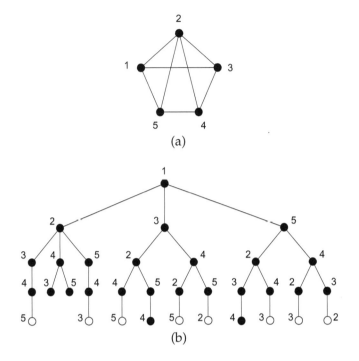

Figure 4.8: (a) A simple graph and (b) all paths originating in vertex 1.

neighbors of u reachable after traversing the path $[v_1, u]$, but excluding v_1. Similarly, for any set $R([v_1, v_2])$ and vertex $u \in R([v_1, v_2])$, we can construct the set $R([v_1, v_2, u])$ consisting of the neighbors of u, excluding v_1 and v_2[1]. In general, we have

$$R([v_1, v_2, \ldots, v_k]) = N(v_k) \backslash \{v_1, \ldots, v_{k-1}\}$$

To illustrate, consider the simple graph in Figure 4.8 and the exploration of all paths originating in vertex 1. In this example, $R([1,3]) = N(3)\backslash\{1\}$, which is equal to the set $\{2,4\}$. Likewise, $R([1,2,4]) = \{3,5\}$. The vertices in Figure 4.8(b) that are colored white are adjacent to vertex 1, meaning that we can complete a Hamilton cycle. One such cycle is $[1, 3, 2, 4, 5, 1]$.

The whole idea is that we continue constructing a set $R([v_1, \ldots, v_k])$ until it becomes empty for some k. Of course, this will be the case for any $k \geq |V(G)|$ as we will have inspected all vertices by then. On the other hand, it is possible that for $k < |V(G)|$ a set already becomes empty, as is the

---

[1] Remember that $V \backslash W$ contains those elements of V that are not also in W. This means that it also excludes the elements that are in W but not in V

4.2. HAMILTON CYCLES

case with $R([1, 2, 4, 3])$ and $R([1, 2, 4, 5])$ in Figure 4.8. When a set becomes empty, we consider only the ones with $k = |V(G)|$ and check whether $v_n \in N(v_1)$. If so, we will have discovered a Hamilton cycle.

Exhaustively enumerating all Hamilton cycles can work only for small graphs. When graphs grow even beyond something like 10 or 15 vertices, other approaches are needed. A relatively simple one that fits into a trial-and-error approach is the following, due to Posa [1976] and described in detail by Vandegriend [1998]. This algorithm makes use of what is known as a **rotational transformation**, which is sketched in Figure 4.9. The idea is that once we have a path $[v_1, v_2, \ldots, v_{j-1}, v_j, \ldots, v_{k-1}, v_k]$ and a "shortcut" by means of an edge $\langle v_1, v_j \rangle$ that we consider exploring an alternative path $[v_{j-1}, \ldots, v_2, v_1, v_j, \ldots, v_{k-1}, v_k]$.

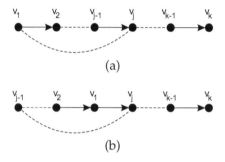

Figure 4.9: Rotational transformation by which the original path (a) is transformed to another one (b) after finding the edge $\langle v_1, v_j \rangle$.

Algorithm 4.3 (Posa): *Consider a graph G and let $u \in V(G)$ be a randomly selected vertex. This vertex forms the first vertex of a path P that is expanded as follows. Let last(P) denote the last vertex of P. Note that initially, last(P) = u.*

1. *Randomly select a neighboring vertex $v \in N(last(P))$, such that (1) preferably, v does not lie on P, and (2) if $v \in V(P)$, then v has not been previously selected as neighbor of a last vertex before. If no such vertex exists, stop.*

2. *If $v \notin V(P)$, set $P \leftarrow P + \langle last(P), v \rangle$, i.e., expand P with the edge $e = \langle last(P), v \rangle$.[2]*

3. *If $v \in V(P)$ then apply a rotational transformation of P using $\langle last(P), v \rangle$, leading to path P^* with a new last vertex $last(P^*)$. If $last(P^*)$ has not yet been the last vertex for paths of the current length, set $P \leftarrow P^*$.*

---

[2]Formally, this means considering the induced graph $G[E(P) \cup \{e\}]$

4. If in the possibly modified version of P we now have that $V(P) = V(G)$, check if $\langle u, last(P)\rangle \in E(G)$. If so, we have found a Hamilton cycle. Otherwise, continue with step 1.

The working of this algorithm is best illustrated through an example. Consider the graph G shown in Figure 4.10 and assume we have already constructed path P as also shown. This construction comes from simply applying the preference rule, by which we attempt to add new vertices until that is no longer possible. We now have the path

$$P = [1, 2, 3, 4, 5, 6]$$

At that point, we can select only from vertices that already lie on P. Assume we randomly selected vertex 4. We then apply a rotational transformation using edge $\langle 6, 4 \rangle$, meaning that after visiting vertex 4 we continue with vertex 6 from where we continue along the original path, but in reversed order. This leads to

$$P' = [1, 2, 3, 4, 6, 5]$$

from which we then should select vertex 7 resulting in path P'' shown in Figure 4.10(c). Unfortunately, there is no edge $\langle 1, 7 \rangle$, so that we continue with step 1 of Posa's algorithm. Assume we select vertex 2. A rotational transformation then yields that we continue with vertex 7 after visiting vertex 2, to subsequently walk the (2,7)-segment of P'' but in the reversed direction, yielding

$$P''' = [1, 2, 7, 5, 6, 4, 3]$$

Because $\langle 1, 3 \rangle \in E(G)$, we have just found a Hamilton cycle, completing the algorithm.

Optimal Hamilton cycles

Finding a Hamilton cycle can already be a computationally hard problem; finding the *best* Hamilton cycle is even more difficult. Best in this context is defined on a weighted graph in which each edge e has a nonnegative weight $w(e)$. We are now seeking Hamilton cycles with minimal weight, i.e., $\sum_{e \in E(C)} w(e)$ should be minimal among all Hamilton cycles.

Finding an optimal Hamilton cycle becomes much easier if we can assume that a graph is complete. In many practical situations, this is actually a reasonable assumption, as we will explain shortly. A simple approach toward tackling this problem is to first construct a trivial Hamilton cycle, and then to subsequently try to modify that cycle such that its total weight reduces. In the case of a weighted complete graph with vertices $v_1, v_2, \ldots, v_n$, we can start with the Hamilton cycle $C = [v_1, v_2, \ldots, v_n, v_1]$. This cycle

4.2. HAMILTON CYCLES

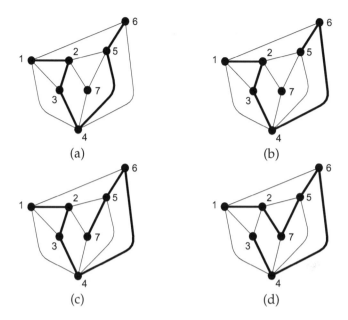

Figure 4.10: Illustration of Posa's algorithm starting from an initial path (a), and applying rotational transformation after selecting vertex 4 (b), adding vertex 7 (c) and selecting vertex 2 followed by a rotational transformation (d).

can be modified by deleting the edges $\langle v_i, v_{i+1}\rangle$ and $\langle v_j, v_{j+1}\rangle$ and replacing them by edges $\langle v_i, v_j\rangle$ and $\langle v_{i+1}, v_{j+1}\rangle$, as shown in Figure 4.11. If

$$w(\langle v_i, v_j\rangle) + w(\langle v_{i+1}, v_{j+1}\rangle) < w(\langle v_i, v_{i+1}\rangle) + w(\langle v_j, v_{j+1}\rangle)$$

we will have found a better Hamilton cycle than C.

As said, assuming that we're dealing with a complete graph is a reasonable assumption in many practical cases, such as finding the optimal route for a traveling salesman. In that case, we are considering n locations connected in some geographical network. We are seeking a closed route such that every location is visited exactly once. This situation can be modeled as a weighted graph with n vertices, where two vertices are joined only if there is a connection between the two in the real network. The corresponding edge has a weight that reflects the real-world distance between its two end points. We can also model the network through a complete graph in which an edge has an extraordinary high weight whenever its two end points are not connected in the real world. Clearly, an optimal Hamilton cycle will never include such an edge, for which reason it can't hurt to include it when representing the geographical network as a graph.

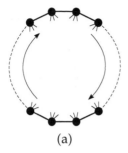 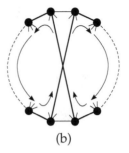

(a)   (b)

Figure 4.11: Modifying a Hamilton cycle in a complete graph (a) to a possibly better one (b).

Exact solutions for the traveling salesman problem have been found for very large networks, including the Swedish road map (with over 24,000) cities. Impressive is also the near-optimal solution for China, comprising over 71,000 cities and provably to close as 0.0024% of the optimal[3].

Note 4.8 (More information)
If the traveling salesman problem is so computationally difficult, how could one ever know that a solution is the best one, or otherwise so close to the best one? The trick is not to try to actually find the best solution, but to estimate the length of the best solution. More specifically, we can try to compute what is known as a **lower bound**: the lowest value that is known that no Hamilton cycle in a given graph can ever reach. If we assume that the minimal weight of an edge is equal to 1, then a trivial lower bound is also 1. In fact, for any simple, connected graph with n vertices and minimal edge weight w, it should be clear that no Hamilton cycle will have a weight less than $n \cdot w$.

It is not hard to imagine that we can generally come to much better estimations of a lower bounds, although these do require some mathematics that are beyond the level of most undergraduate courses. For this reason, we shall not discuss them any further. However, you may ask how you can actually *prove* that a solution is optimal. The answer is quite simple: assume some approach finds a Hamilton cycle and at the same time using completely different methods, we happen to know that a lower bound is equal to LW. In other words, we have shown that for all Hamilton cycles C, we know that for weight $w(C)$ of C we have $w(C) \geq LW$, then obviously, if we find a C for which $w(C) = LW$, that cycle must be optimal. Note that this just tells us that only *one* solution has been found.

---

[3]See also http://www.tsp.gatech.edu/world/countries.html

4.2. HAMILTON CYCLES

Another issue is that in real networks links may not be symmetric: the distance from A to B may be different than the one from B to A. For example, many end users are connected to the Internet through what is known as an ADSL subscription. Such a subscription is characterized by the fact that data that is received by the end user is transmitted at a higher rate than data that is sent by that user. In such situations, we need to model the network as a weighted directed graph, and subsequently find an optimal directed Hamilton cycle.

The question that comes to mind is how we can use techniques for finding (optimal) Hamilton cycles in undirected graphs for the directed case. The answer to this question lies in transforming weighted directed graphs to an equivalent weighted undirected form. To this end, we proceed as follows. Consider a directed Hamiltonian graph D with n. We construct a undirected Hamiltonian graph $\hat{D}$ with $3n$ vertices by representing each vertex $v \in V(D)$ by the triplet (v_{in}, v, v_{out}), as shown in Figure 4.12.

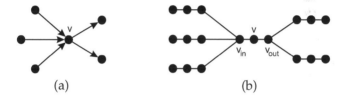

Figure 4.12: Transforming (a) a directed Hamiltonian graph to (b) an equivalent undirected graph.

In the case we are dealing with weights, the weight of an arc $\langle \overrightarrow{u,v} \rangle$ is represented by the same weight on the edge $\langle u_{out}, v_{in} \rangle$, whereas all other edges have weight 0. We now have:

Theorem 4.8: *A directed graph D is Hamiltonian if and only if its transformed undirected version $\hat{D}$ is Hamiltonian.*

Proof. First assume that D is Hamiltonian and let $C = [v^1, v^2, \ldots, v^n, v^1]$ be a Hamilton cycle. Clearly, the cycle

$$\hat{C} = [v^1, v^1_{out}, v^2_{in}, v^2, v^2_{out}, \ldots, v^n_{in}, v^n, v^n_{out}, v^1_{in}, v^1]$$

is a Hamilton cycle in $\hat{D}$.

Conversely, consider a Hamilton cycle $\hat{C}$ in $\hat{D}$. Obviously, for each vertex $v^k \in V(\hat{D})$, $\hat{C}$ contains the edges $\langle v^k_{in}, v^k \rangle$ and $\langle v^k, v^k_{out} \rangle$, for otherwise it would be impossible to have visited vertex v^k. For this reason, $\hat{C}$ corresponds to a unique directed Hamilton cycle C in D. □

In the case of the directed equivalent of the traveling salesman problem, we need to assume that the corresponding weighted directed graph D is strongly connected. In that case, when there is no direct connection from location A to B (e.g., because we are dealing with one-way streets), we can still be sure that in the transformed *complete* graph with $3n$ vertices, there will be an (A, B)-path corresponding to a directed (A, B)-path in D.

Chapter 5

Trees

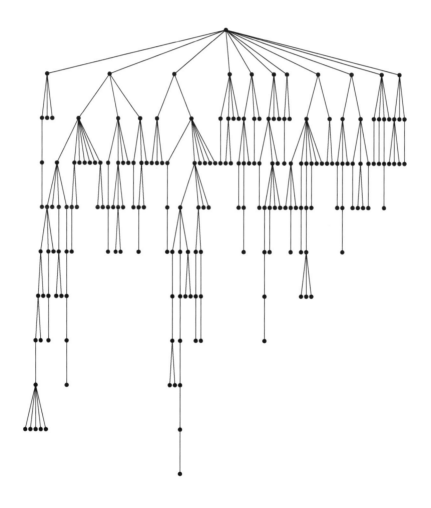

In the previous chapter we occasionally came across graphs lacking cycles. Such graphs are also known as **trees**. Trees form a special type of graph and are important to study if only for their common and widespread use in diverse fields of practice and science. In this chapter we shall take a closer look at trees, starting with presenting various applications. We will then take a look at some formal issues, after which we concentrate on the construction of optimal trees that can be used span an entire graph, or for finding shortest paths in a weighted directed or undirected graph.

5.1 Background

Before we delve into various details and formalities, let us first consider why trees receive so much attention. There are different fields in which trees are extensively applied. Below, we just mention two of the more salient ones.

Trees in transportation networks

A compelling example of the application of trees is in transportation networks. Typical examples of such networks include communication networks and traffic networks, but also networks related to logistics such as those reflecting the transportation of goods. In many cases, we need to solve the problem of minimizing transportation costs from a source to multiple destinations (or *vice versa*). In practice, this boils down to finding the cheapest paths in a network. We already came across this problem when we discussed Dijkstra's algorithm in Chapter 3. In that case, finding the cheapest paths involved building a tree rooted at a particular vertex u and constructing all cheapest (v, u)-paths from other vertices v. We will return to finding cheapest paths later in this chapter.

A variation of this problem is that of setting up a communication infrastructure between a collection of nodes but such that the total costs are minimized. For example, the nodes could be towns, the infrastructure is a railway network, and the costs between two nodes corresponds to the distance that needs to be covered. This example is also known as the **connector problem**.

The connector problem has practical instances in communication networks. Consider the delivery of video streams over the Internet. A famous project that aimed at efficiently providing the facilities for such a service was the MBone [Eriksson, 1994; Macedonia and Brutzman, 1994], an abbreviation for Multicast Backbone. This network consisted of many so-called MBone routers, which were just normal computers spread across the Internet. Important was the fact that two such routers would maintain a permanent connection that could be used for streaming audio and video packets.

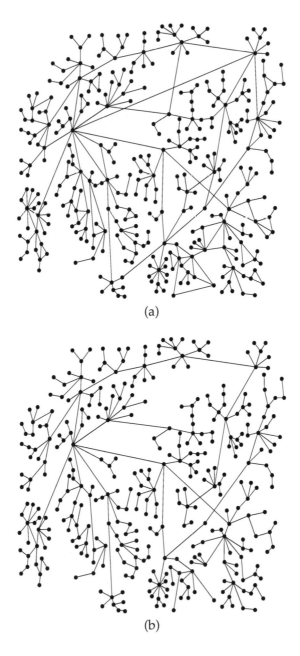

Figure 5.1: (a) A map of the MBone overlay network as of July 1993, and (b) a spanning tree of that network.

5.1. BACKGROUND

In graph terminology, the routers formed the vertices, whereas the connections would form edges. The result is what is known as an **overlay network**, a concept we shall also come across when discussing peer-to-peer networks in Chapter 8.

To give an idea, Figure 5.1(a) shows a map of the MBone network as of July 1993. At that point, there where approximately 400 MBone routers maintaining the connections as shown. It is worth mentioning that we are already dealing with a network that has only a few cycles. Figure 5.1(b) shows a **spanning tree** of the MBone, that is, an acyclic connected subgraph of the MBone with the same set of vertices. How we can compute such a tree is discussed below. With a spanning tree in place, there is no further need to set up routes. In the case of the MBone, nodes could join or leave *groups*, with each group essentially representing those nodes that were interested in the same video stream. Members of the same group were subsequently connected to each other by means of a spanning tree. Note that in this case, a spanning tree needed to reach out only to the nodes in the same group, and not necessarily to all nodes comprising the MBone.

Trees as data structures

Trees are also used extensively to organize data in computer systems. In particular, they appear as so-called **rooted trees**, which is a tree with a single vertex designated as the root. To given an example of how trees can be used to represent data, consider the following well-known arithmetic expression describing one solution (if it exists) of the quadratic equation $ax^2 + bx + c$:

$$x = \frac{-b + \sqrt{b^2 - 4ac}}{2a}$$

Computers need to process such expressions, to which end they first need to be stored. This can be done conveniently in the form of the rooted tree shown in Figure 5.2. The tree contains two types of nodes. The **leaf nodes**, which are the ones having degree 1 forming the "lowest level" nodes, contain the *variables* or *constants*. In our example, we have one variable, namely x, and three constants, a, b, and c. The other, **intermediate nodes**, represent *operations*.

The link between the original expression and the tree may be better understood if we rewrite the expression as

$$x = (-b + \text{sqrt}(b*b - 4*(a*c)))/(2*a)$$

where we now use the function *sqrt(y)* as an equivalent notation for $\sqrt{y}$. Note how each operator has either one or two *descendants*, depending on

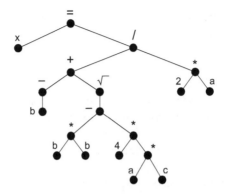

Figure 5.2: The representation of an arithmetic expression as a tree.

whether we are dealing with a *unary* operator (which operates on one variable or constant), or a *binary* operator (which takes two arguments).

Note 5.1 (More information)
In fact, we can replace each of the other operations with functions like *sqrt* as follows:

operation	function	type
=	eq	binary
+	sum	binary
−	min	binary
*	mul	binary
/	div	binary
−	neg	unary
√	sqrt	unary

As said, we make a distinction between *binary* and *unary* operations, where it should be noted that the operation "−" is used in two different forms. Note also that *sqrt* is indeed a unary operation. With these functions, we can rewrite our original expression as:

eq(x,div(add(neg(b),sqrt(min(mul(b,b),mul(4,mul(a,b))))),mul(2,a)))

What has happened in comparison to the original expression, is that we have switched from what is known as an **infix notation** to a **prefix notation**. In the former, operators are placed *between* variables and constants, whereas with the latter they are placed *in front* of them. To a computer it makes no difference. The only thing that does matter is the organization of the rooted tree as given in Figure 5.2, as this tree is an unambiguous representation of the expression.

5.1. BACKGROUND

The common terminology for rooted trees that are used for data structures is to say that every node has one or more **descendants**. Likewise, each node except the root node is said to have a **parent**. Note also that each node u having k descendants is the root of k subtrees, each subtree in turn rooted by a respective descendant of u. A special case is a **binary tree** in which there are exactly two descendants for each intermediate node.

Binary trees come in handy when we need to quickly look up elements in a finite ordered set. An ordered set $S = \{x_1, x_2, \ldots, x_n\}$, has the property that $x_i < x_j$ if $i < j$. As an example, any finite subset of natural numbers forms an ordered set. Consider the set

$$S = \{3, 6, 8, 12, 15, 20, 21, 27, 32, 33, 34, 45, 49, 51, 56, 60, 61\}.$$

This set consists of 17 elements. To represent it as a binary tree, each element will be represented by a node. We demand for each node x that all elements in the left subtree represent values that are less than x, and all elements in the right subtree have larger values. Leaf nodes store no value; they are simply added for convenience as we discuss next.

Figure 5.3 shows the representation of our set S as a binary tree. Now suppose that we wish to look up the element $x = 16$, which is not contained in S. In this case, we start at the root node, which has value 27, continue in its left subtree reaching 12. If $x = 16$ is contained in S, it must be stored in the right subtree, which brings us to node 20, and from there on to 15, where we can stop the traversal, now knowing that 16 is not in S.

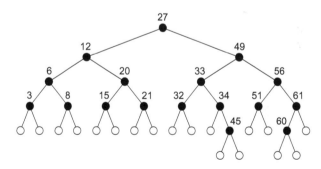

Figure 5.3: The representation of a set of natural numbers as a binary tree.

Now compare this lookup operation with the situation that we had represented S as a *list*. In that case, we would start from the first element and subsequently move through the list until reaching the first element larger than 16. On *average*, a lookup operation would require inspecting $|S|/2$ elements if S had been represented as a list. In the case of using binary trees, one can show that the number of operations are approximately $\log_2(|S|)$, which is considerably less as S becomes larger. Further details on how to use trees for representing data in computers can be found in [Goodrich and Tamassia, 2002].

5.2 Fundamentals

Before discussing various tree-related algorithms, let's first consider a few characteristic features of trees. We start with the following observation:

Theorem 5.1: *For any connected (simple) graph G with n vertices and m edges, $n \leq m + 1$.*

Proof. The proof proceeds by induction on the number of edges m. Clearly, if $m = 1$, we necessarily have $n = 2$ so that the theorem is true. Now assume the theorem is true for all graphs with fewer than k edges and consider a graph G with exactly k edges and n vertices.

Suppose that G contains a cycle C. In that case, choose an arbitrary edge $e \in E(C)$ and construct the induced subgraph $G^* = G - e$. Because e was lying on the cycle C, G^* will still be connected, meaning that $n = |V(G^*)| \leq |E(G^*)| + 1 = (k-1) + 1 = k$. But in that case, we certainly have that $n \leq k + 1$.

If G does not contain a cycle, find a longest path P in G. Let u and w be the end points of P. Note that because G is smple, the degree of each these nodes must be 1, for otherwise P could not have been a longest path. Now consider the induced subgraph $G^* = G - u$. Clearly, G^* is connected and we have $|V(G^*)| = n - 1$ and $E(G^*) = k - 1$. By induction, we thus also have that $n - 1 \leq (k - 1) + 1 = k$, and thus $n \leq k + 1$, completing our proof. □

Note 5.2 (Proof techniques)
Again, we have encountered a proof by induction. Note that the approach we have taken is common to many such proofs. After having proven that the theorem holds for an initial, generally almost trivial case, we proceed with assuming that the theorem holds up until and including the case that $m = k$. We then consider a situation with $k + 1$ edges. In our attempt to prove the theorem, we try to *reduce* the new graph to one with at least one edge less, knowing that in that case we can assume the theorem holds. This brings us to a new starting situation from where on we need to show that the theorem also holds in the original situation with $k + 1$ edges.

Furthermore, note also that we have combined a proof by induction with extremality by looking at a longest path P, from which we then remove an edge

5.2. FUNDAMENTALS

> "at the extreme." As we stated before, it is important to fully understand these proofs, as they enforce you to understand details and techniques that are common to many graph-theoretical problems.

Note that, because the theorem holds for simple graphs, it certainly holds for nonsimple graphs as well. It should now come as no real surprise that trees obey the following property:

Theorem 5.2: *For any tree T with n vertices and m edges, $n = m + 1$.*

Note that we already proved this theorem in Chapter 2 (Lemma 2.1 on page 51). We leave it as an exercise to the reader to provide an alternative proof, based on the proof of Theorem 5.1. Interestingly, the implication formulated in the previous proof also holds in the opposite direction:

Theorem 5.3: *A connected graph G with n vertices and m edges for which $n = m + 1$, is a tree.*

Proof. We prove the theorem by contradiction. To this end, assume G is not a tree, i.e., it contains a cycle C. Let edge $e \in E(C)$. Obviously, the induced subgraph $G - e$ is still connected, but with one edge less than G. From Theorem 5.1 we know that $|V(G-e)| \leq |E(G-e)| + 1$. With $|V(G-e)| = n$ and $|E(G-e)| = m - 1$, we thus have that $n \leq (m-1) + 1 = m$. However, we assumed that $n = m + 1$, which contradicts that $n \leq m$. Hence, our initial assumption, namely that G is not a tree, was false. □

Let us proceed with another important characterization of trees:

Theorem 5.4: *A graph G is a tree if and only if there exists exactly one path between every two vertices u and v.*

Proof. Recall that the phrase "if and only if" means that we need to prove two things: (1) If G is a tree then there exists a unique path between every two vertices and (2) if there exists a unique path between every two vertices, then G is tree.

(1) Let G be a tree and let u and v be two distinct vertices. Because G is connected, there exists a (u,v)-path P. Assume there is another, distinct (u,v)-path Q. Let x be the last vertex on P that is also on Q when traversing P starting from u. In other words, the next vertex following x will be different for P and for Q, as shown in Figure 5.4. Likewise, let y be the first vertex succeeding x that is common to both

Figure 5.4: The construction of a cycle based on two distinct (u,v)-paths.

P and Q again. We have now identified a cycle in G, contradicting that G was a tree.

(2) Now assume that G is not a tree. Note that because there is a path between every two vertices, G is connected. If G is not a tree, there must be a cycle $C = [v_1, v_2, \ldots, v_n = v_1]$. Clearly, for every two distinct vertices v_i and v_j ($i < j$) on C we have also have two distinct (v_i, v_j)-paths: $P_1 = [v_i, v_{i+1}, \ldots, v_{j-1}, v_j]$ and $P_2 = [v_i, v_{i-1}, \ldots, v_2, v_1 = v_n, v_{n-1}, \ldots, v_{j+1}, v_j]$, which contradicts the uniqueness of paths. □

Before we provide another characterization of trees, we prove the following, intuitively simple theorem:

Theorem 5.5: *An edge e of a graph G is a cut edge if and only if e is not part of any cycle of G.*

Proof. Again, we need to prove two things: (1) If e is not part of any cycle, then e is a cut edge, and (2) if e is a cut edge, it cannot be part of any cycle of G.

(1) By contradiction: assume that $e = \langle u, v \rangle$ is not a cut edge (and not part of any cycle). If e is not a cut edge, then u and v must still be in the same component of $G - e$. This implies that there is a (u,v)-path P in $G - e$ connecting u and v. However, this also means that $P + e$ is a cycle in G, which violates our assumption.

(2) Again, by contradiction: let $e = \langle u, v \rangle$ be a cut edge of G and let x and y be two vertices in different components of $G - e$. Because there is an (x,y)-path P in G connecting x and y, we necessarily have that e is part of P. Assume that u precedes v when traversing P from x to y. Let P_1 be the (x,u)-path part of P and P_2 the (v,y)-path that is part of P. If e were part of a cycle C, then u and v would be connected in $G - e$ through the path $C - e$. Let u^* be the first vertex common to P_1 and $C - e$ when traversing P_1 from x. Likewise, let v^* be the first vertex common to P_2 and $C - e$ when traversing P_2 from y. Let $a \xrightarrow{Q} b$ denote that part of path Q that connects vertex a to b. Clearly, the path

$x \xrightarrow{P_1} u^* \xrightarrow{C-e} v^* \xrightarrow{P_2} y$ connects x and y in $G - e$, contradicting that e was a cut edge. Hence, e cannot be part of any cycle. □

> **Note 5.3** (Study tip)
> For this proof, it is helpful to draw a diagram for case (2). We deliberately leave this is an exercise, anticipating that you will gain more insight in the construction of the proof.

With this result, we can now easily prove the following characterization:

Theorem 5.6: *A connected graph G is a tree if and only if every edge is a cut edge.*

Proof. Again we need to prove two things: (1) If G is a tree then every edge is a cut edge, and (2) if every edge is a cut edge, then G is a tree.

(1) Let G be a tree and e an edge of G. Because G contains no cycles, e is also not contained in any cycle, meaning that it must be a cut edge.

(2) Assume G contains a cycle C. However, we now know that none of the edges of C can be a cut edge, which means that not every edge in G is a cut edge, contradicting our starting-point. □

To summarize, we have described the following equivalent statements (1–5) for a graph G with n vertices and m edges:

1. G is a tree, that is, it is connected and acyclic.
2. G is connected with $n = m + 1$.
3. G is acyclic with $n = m + 1$.
4. There exists a unique path between every two vertices.
5. G is connected and every edge is a cut edge.
6. G does not contain a cycle and adding a single edge creates a unique one in G.

We leave the proof of the last statement as an exercise to the reader. These theorems together provide a handful of characterizations of trees which will show to be useful when determining properties of various networks. In what follows, we shall concentrate on constructing specific trees as subgraphs of networks.

5.3 Spanning trees

As we've mentioned, a spanning tree of a connected graph G is an acyclic connected subgraph of G containing all of G's vertices. It is not difficult to see that every connected graph G has a spanning tree. Let T be a spanning subgraph of G with a minimal number of edges. Clearly, every edge e in T is a cut edge of T, for otherwise T would not be minimal. Hence, from Theorem 5.6 we know that T is necessarily a tree.

More interesting than just noticing that a connected graph G has a spanning tree, is finding a minimal spanning tree in a *weighted* connected graph G. In other words, our goal is to find a spanning tree T with minimal weight among all spanning trees of G. Recall that the weight of a subgraph H is simply defined as the total sum of the weights of H's edges: $w(H) = \sum_{e \in E(H)} w(e)$. A famous algorithm that efficiently constructs such a tree was designed by Kruskal [1956].

> **Note 5.4** (More information)
> By the way, Kruskal was not the first to have devised a solution to the minimal spanning tree problem, which is generally attributed to the Czech mathematician Otakar Borůvka back in 1926, even before graph theory was "invented." It is uncertain whether there was already a solution as early as 1909. See also Graham and Hell [1985].

Algorithm 5.1 (Kruskal): *Consider a weighted graph G where each edge e has been assigned a real-valued weight $w(e) \in \mathbb{R}$. Choose an edge e_1 with minimal weight.*

1. *Suppose that edges $E_k = \{e_1, e_2, \ldots, e_k\}$ have been chosen so far. Choose a next edge e_{k+1} from $E(G) \backslash E_k$ such that the following two conditions are met:*

   (1) *The induced subgraph $G_{k+1} = G[\{e_1, e_2, \ldots, e_k, e_{k+1}\}]$ is acyclic (note that we are not demanding that G_{k+1} is also connected).*

   (2) *The weight $w(e_{k+1})$ is minimal, i.e., for all $e \in E(G) \backslash E_k$, we know that $w(e) \geq w(e_{k+1})$.*

2. *Stop when there is no more edge to select in the previous step.*

To get an impression how Kruskal's algorithm works, Figure 5.5 shows a weighted complete graph on eight vertices. The edges have been assigned random weights. If we sort the edges by weight, we can see more clearly how the algorithm works, as expressed in Figure 5.6. The resulting tree has a total weight of 190.

5.3. SPANNING TREES

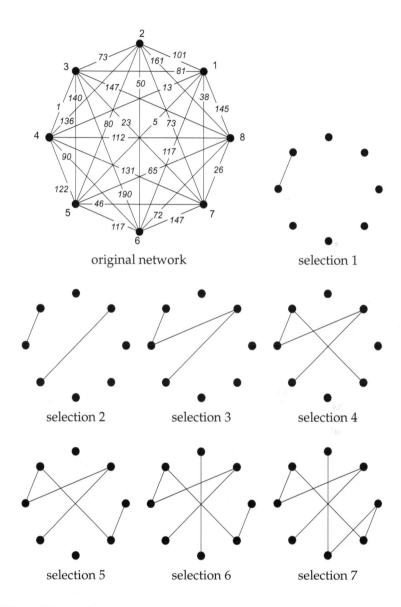

Figure 5.5: Applying Kruskal's algorithm to finding a minimal spanning tree.

Edge	Weight	Comment
$\langle 3,4 \rangle$	1	Selection 1: added
$\langle 1,5 \rangle$	5	Selection 2: added
$\langle 1,4 \rangle$	13	Selection 3: added
$\langle 3,7 \rangle$	23	Selection 4: added
$\langle 7,8 \rangle$	26	Selection 5: added
$\langle 1,7 \rangle$	38	Cannot add: creates a cycle $[1,7,3,4,1]$
$\langle 5,7 \rangle$	46	Cannot add: creates a cycle $[1,5,7,3,4,1]$
$\langle 2,6 \rangle$	50	Selection 6: added
$\langle 5,8 \rangle$	65	Cannot add: creates a cycle $[1,5,8,7,3,4,1]$
$\langle 6,8 \rangle$	72	Selection 7: added, completing the tree

Figure 5.6: The evaluation of Kruskal's algorithm on the graph from Figure 5.5.

From this table we can also see that the algorithm is relatively efficient: we simply need to sort all edges by their weight and subsequently inspect each edge starting from the one with the lowest weight. Of course, things get somewhat more complicated when cycles need to be detected, yet even then no big issues arise. For example, when inspecting an edge $e = \langle u,v \rangle$, we need merely check whether e is joining two vertices that are already in the same component, for in that case there would be a (u,v)-path P, thus leading to the cycle $P + e$. If the two end points are not in the same component, we can safely add e. Identifying the components in a graph is relatively easy and is left as an exercise. This also means that we can easily check whether e connects two different components, which, together with the fact that it has minimal weight of the remaining edges, is enough to add it to the subgraph constructed so far.

What remains is to show that Kruskal's algorithm is *correct* in the sense that it indeed provides us with an optimal spanning tree. We formulate this as the following theorem (see also Bondy and Murty [1976]):

Theorem 5.7: *Consider a weighted graph G with n vertices. Any spanning tree $T_{Kruskal}$ of G constructed by Kruskal's algorithm has minimal weight.*

Proof. This is typically a theorem that we should prove by contradiction. To this end, consider a spanning tree $T \neq T_{Kruskal}$. Let $\iota(T)$ denote the smallest index i such that when adding edges to $T_{Kruskal}$ according to Kruskal's algorithm, edge $e_i \notin E(T)$. Now assume that $T_{Kruskal}$ is not optimal and let T be a spanning tree with maximal $\iota(T)$. In other words, for any other spanning tree $T' \neq T_{Kruskal}$, we have that T contains at least as many edges from $T_{Kruskal}$ as T'. We will now construct an optimal spanning tree $\hat{T}$ for which $\iota(\hat{T}) > \iota(T)$, thus contradicting our choice of T and our assumption that $T_{Kruskal}$ is not optimal.

Suppose that $\iota(T) = k$, meaning that all edges $e_1, e_2, \ldots, e_{k-1}$ are both edges in T as well as in $T_{Kruskal}$. It can be easily seen that the graph $T + e_k$ contains a unique cycle C. Let $\hat{e}$ be an edge of C such that $\hat{e} \notin E(T_{Kruskal})$, but $\hat{e} \in E(T)$. Because $\hat{e}$ lies on C, it cannot be a cut edge of $T + e_k$. This also means that $\hat{T} = (T + e_k) - \hat{e}$ is also a connected subgraph of G, and thus also a spanning tree. Note that the total weight $w(\hat{T})$ of $\hat{T}$ is equal to

$$w(\hat{T}) = w(T) + w(e_k) - w(\hat{e})$$

An important observation is that edge e_k was chosen to be one with minimal weight that kept the constructed subgraph up to that point acyclic. Clearly, the graph induced by edges $e_1, e_2, \ldots, e_{k-1}, \hat{e}$ is also acyclic, so that we must conclude that $w(\hat{e}) \geq w(e_k)$, and hence, $w(\hat{T}) \leq w(T)$. This can only mean that $\hat{T}$ is also optimal. However, because $e_k \in E(\hat{T})$, we know that $\iota(\hat{T}) > \iota(T)$, which contradicts our choice of T, namely as the tree with the largest value for ι. □

5.4 Routing in communication networks

Trees play a prominent role in communication networks, whose main job is ensuring that messages are sent from their source to their intended destination(s), also referred to as **message routing**. How message routing is accomplished is laid down in a **routing protocol**: a collection of specifications describing exactly what to do when a node in a network receives a message from source A that is destined for node B. In general, a node in a communication network can be viewed as consisting of several **interfaces**, where each interface connects that node to exactly one other node in the network. In this way, we can represent a communication network as a graph with nodes as vertices and links between two nodes as edges. An interface is actually the end point of a link, and its representation coincides with the vertex representing the node to which that link is attached.

A node usually maintains a **routing table**. Each row in this table specifies to which interface a message should be forwarded, given its source and destination, and optionally also the interface through which it arrived. An important function of a routing protocol is constructing these tables. This is exactly what we established when discussing Dijkstra's shortest path algorithm in Section 3.2: each node maintained exact information on the next closest node to which a message should be routed, including how far a message would still need to travel.

Crucial for routing is that messages are not endlessly forwarded. Technically, this means that for every destination u messages should follow paths in a spanning tree that is said to be *rooted* at u, hence called a **rooted tree**. In particular, and analogous to what we also mentioned in Section 3.2, rooted in this case means that we are interested only in (v,u)-paths, where v indicates the source node. With u being the destination node, such a rooted tree is also called a **sink tree** for u.

Dijkstra's algorithm

The issue for routing protocols is to construct these sink trees, one for every node in the network. A famous one is Dijkstra's algorithm, which we already discussed in Chapter 3. There, we illustrated how the algorithm works for undirected graphs. It is not difficult to see that the algorithm also works for directed graphs, and, in fact, that it can be easily formulated to construct sink trees. The only restriction we demand is that the weight associated with an arc is nonnegative. Dijkstra's solution for constructing optimal routes is so important that it is worthwhile also considering its counterpart for directed graphs. For example, it is widely deployed in communication networks where it is known as a **link-state routing protocol** (see, for example, Moy [1995]). The following description of the algorithm is nearly identical to the one given in Chapter 3, except that we now construct paths *to* the root vertex u.

Algorithm 5.2 (Dijkstra, sink tree construction): *Consider a directed, weighted graph D where weights are nonnegative, and a vertex $u \in V(D)$. We introduce the following sets and labels:*

- *Let $S_t(u)$ be the set of vertices from which a shortest path to vertex u has been found after step t.*
- *Each vertex v is assigned a label $\mathbf{L}(v) \stackrel{\text{def}}{=} \big(L_1(v), L_2(v)\big)$, in which $L_1(v)$ is the vertex succeeding v in the shortest (v,u)-path found so far, and $L_2(v)$ the total weight of that path.*
- *Let $R_t(u) \stackrel{\text{def}}{=} S_t(u) \cup_{v \in S_t(u)} N_{in}(v)$, with $N_{in}(v)$ denoting the set of in-neighbors of v. In other words, $R_t(u)$ consists of all vertices in $S_t(u)$ and the vertices from where $S_t(u)$ can be reached through an arc.*

1. *Initialize $t \leftarrow 0$ and $S_0(u) \leftarrow \{u\}$. Furthermore, for all $v \in V(G)$:*

$$\mathbf{L}(v) \leftarrow \begin{cases} (u,0) & \text{if } v = u \\ (-,\infty) & \text{otherwise} \end{cases}$$

2. For each vertex $y \in R_t(u) \backslash S_t(u)$, consider the vertices $N'_{out}(y)$ that are out-neighbors of y that lie in $S_t(u)$, i.e., $N'_{out}(y) \stackrel{\text{def}}{=} N_{out}(y) \cap S_t(u)$. Select $x \in N'_{out}(y)$ for which $L_2(x) + w(\langle \overrightarrow{y,x} \rangle)$ is minimal. Set $\mathbf{L}(y) \leftarrow (x, L_2(x) + w(e))$.

3. Let $z \in R_t(u) \backslash S_t(u)$ for which $L_2(z)$ is minimal. Set $S_{t+1}(u) \leftarrow S_t(u) \cup \{z\}$. If $S_{t+1}(u) = V(G)$, stop. Otherwise, $t \leftarrow t+1$, compute $R_t(u)$ again and repeat the previous step.

To illustrate this algorithm, let us reconsider the graph from Figure 3.4, but now with its edges be directed, as shown in Figure 5.7. What we see is that we can apply the same steps, but, of course, because the graph is now directed, we obtain a different (directed) tree rooted at vertex v_0. Again, we can formulate this version of Dijkstra's algorithm in pseudo-code, which is left to the reader.

> **Note 5.5** (Mathematical language)
> Despite our deliberate use of formal notations, by now it should be clear from the mathematical description what the principle behind Dijkstra's algorithm is. Every time we have completed the set $S(u)$, we attempt to expand it by adding a vertex from the next ring of vertices from where $S(u)$ can be reached, and subsequently add the vertex closest to u, as shown in Figure 5.8.
>
>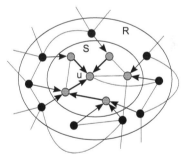
>
> **Figure 5.8:** An illustration of the relation between S and R.
>
> To properly understand algorithms such as the one from Dijkstra, it is important to develop these type of high-level insights. Drawings generally help a lot and force you to translate the mathematical concepts into simpler principles, in turn, assisting in understanding those concepts.

Although Dijkstra's algorithm is relatively simple, it is not obvious that it is also correct. We follow Goodrich and Tamassia [2002] in proving its correctness.

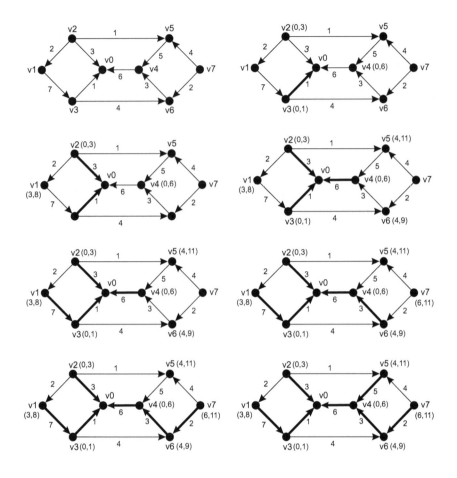

Figure 5.7: Applying Dijkstra's algorithm to construct a sink tree in a weighted directed graph.

Theorem 5.8: *Given a weighted directed graph D. When applying Dijkstra's algorithm to a vertex u, each time a vertex z is added to the set $S_t(u)$, $L_2(z)$ corresponds to the length of a shortest (z,u)-path.*

Proof. By contradiction. Let $d(w,u)$ denote the total weight of an optimal (w,u)-path. Let z be the first vertex that was added to an $S_t(u)$ for which $L_2(z) > d(z,u)$. In other words, up until and including step t we have that for all vertices $v \in S_t(u)$, $L_2(v) = d(v,u)$, but $S_{t+1}(u)$ contains, for the first time a vertex z for which $L_2(z) > d(z,u)$. Because z was selected (after t steps), we know that $L_2(z) < \infty$ and thus that there is a (z,u)-path. In particular, there must be a shortest (z,u)-path, say P. Let y be the last vertex on that path (from z to u) that is not in $S_t(u)$, and x its successor (and thus in $S_t(u)$). By choice of z, we know that $L_2(x) = d(x,u)$, i.e., $L_2(x)$ is equal to the total weight of an optimal (x,u)-path.

When x was selected (say, at step t'), we also evaluated y and possibly adjusted $L_2(y)$ so that the value of $L_2(y)$ is in any case at most $L_2(x) + w(\langle \overrightarrow{y,x} \rangle)$, i.e., $L_2(y) \leq L_2(x) + w(\langle \overrightarrow{y,x} \rangle)$. On the other hand, because y is on the shortest (z,u)-path P, x is the successor of y on P, and $L_2(x) = d(x,u)$, we necessarily have that $L_2(x) + w(\langle \overrightarrow{y,x} \rangle) = d(x,u) + w(\langle \overrightarrow{y,x} \rangle)$ must correspond to the length of a shortest (y,u) path, i.e. $L_2(x) + w(\langle \overrightarrow{y,x} \rangle) = d(y,u)$. However, we have to realize that y was *not* selected to be included in an $S_t(u)$, which can only mean that $L_2(z) \leq L_2(y)$. Because y is on a shortest (z,u)-path, we also have

$$d(z,y) + d(y,u) = d(z,u)$$

and because $d(y,u) \geq 0$, we now have that:

1. $L_2(z) \leq L_2(y)$
2. $L_2(y) = d(y,u)$
3. $d(y,u) \leq d(y,u) + d(z,y)$
4. $d(y,u) + d(z,y) = d(z,u)$

and thus that $L_2(z) \leq d(z,u)$, contradicting our choice of z. Hence, the assumption that there exists a z that was added to $S_t(u)$ with $L_2(z) > d(z,u)$ is false, completing our proof. □

The Bellman-Ford algorithm

An important observation is that in order to execute Dijkstra's algorithm, we need to know exactly what the graph looks like. In other words, we need to know which vertices are adjacent to each other and what the weight of their respective connecting edges are. We say that we need to know the **topology** of the graph. In practice, when a node u in a communication network receives a message intended for node v, it needs to forward that message along the optimal sink tree for v. The same holds for any other incoming message regardless its destination, As a consequence, node u will have to precompute the optimal sink tree for each node in the network. In real networks, we therefore see that the topology of a network is first spread to all nodes in that network (and, of course, on a regular basis because networks change).

Given this situation, one can ask whether it is possible to compute optimal sink trees *without* having to know the topology in advance. In fact, it is actually not necessary that a node needs to know a complete sink tree, as long as it knows to which next node it should forward an incoming message and that this forwarding is done along an optimal sink tree. A solution to this problem was provided by several people, but is generally known as the **Bellman-Ford algorithm**. It was the basis for one of the first widely applied routing protocols in the Internet, but for reasons we briefly discuss below, it has been largely replaced by protocols based on Dijkstra's algorithm.

The protocol can be completely described from the perspective of a node. To this end, we proceed in *rounds* by letting each node v_i compute the optimal path to other nodes based on the information that is available to v_i in that round. Let $d^t(i,j)$ denote the total weight of the optimal (v_i, v_j) path that vertex v_i has found after round t. We denote this total weight as the **routing cost** of getting a message from v_i to v_j. Initially, we have

$$d^0(i,j) \leftarrow \begin{cases} 0 & \text{if } i = j \\ \infty & \text{otherwise} \end{cases}$$

In other words, we let each node initially set the cost to itself to be zero, and the cost to any other node as infinite. We now let v_i adjust its value of d^t_{ij} as follows:

$$d^{t+1}(i,j) \leftarrow \min_{k \in N(v_i)} \{w(v_i, v_k) + d^t(k,j)\}$$

in which $N(v_i)$ is the collection of neighboring nodes of v_i and $w(v_i, v_k)$ the weight of edge $\langle v_i, v_k \rangle$. Note that as soon as d^t_{ij} becomes anything else than infinite, v_i will have discovered a path to v_j. In particular, after the first round, v_i will discover a path to each of its respective neighbors, namely the path consisting of the edge connecting v_i to that neighbor. After two rounds, optimal paths of length 2 will have been discovered, and so on.

In practice, the algorithm is implemented by letting nodes exchange information found in their respective routing tables. Consider the undirected version of Figure 5.7. Each node (which is represented by a vertex of the graph shown in Figure 5.7) will initially know only about itself and no other node. After one round, the routing tables for each node will be as shown in Figure 5.9. We use the notation (d, v) to indicate that a path of cost d has been found, for which messages are to be forwarded to adjacent node v.

Now consider node v_1, who, after one round, has discovered paths to v_2 and v_3. At a certain moment, node v_2 and v_3 will each pass their routing table to v_1. Assume that v_3 was first. In that case, v_1 learns that v_3 has discovered a path to node v_0 at cost 1. Because v_1 has a path to v_3, it has now discovered a path to v_0 at cost 8, for which it need only forward messages to its neighbor v_3. However, as soon as v_2 has passed its routing table to

5.4. ROUTING IN COMMUNICATION NETWORKS

	v_0	v_1	v_2	v_3	v_4	v_5	v_6	v_7
v_0:	$(0,v_0)$		$(3,v_2)$	$(1,v_3)$	$(6,v_4)$			
v_1:		$(0,v_1)$	$(2,v_2)$	$(7,v_3)$				
v_2:	$(3,v_0)$	$(2,v_1)$	$(0,v_2)$			$(1,v_5)$		
v_3:	$(1,v_0)$	$(7,v_1)$		$(0,v_3)$			$(4,v_6)$	
v_4:	$(6,v_0)$				$(0,v_4)$	$(5,v_5)$	$(3,v_6)$	
v_5:			$(1,v_2)$		$(5,v_4)$	$(0,v_5)$		$(4,v_7)$
v_6:				$(4,v_3)$	$(3,v_4)$		$(0,v_6)$	$(2,v_7)$
v_7:						$(4,v_5)$	$(2,v_6)$	$(0,v_7)$

Figure 5.9: The routing tables for the nodes in the undirected version of Figure 5.7, after one round of the Bellman-Ford algorithm.

v_1, the latter will discover a better path to v_0, namely one via v_2 and at total cost $w(v_1, v_2) + d(v_2, v_0) = 2 + 3 = 5$.

Completely analogous, v_0 will eventually pass its routing table to v_2, in which case v_2 will discover a path to v_3 at cost $w(v_2, v_3) + d(v_0, v_3) = 3 + 1 = 4$. It can be readily verified that after the second round, the routing tables will be as shown in Figure 5.10. Note that there are two different paths of equal cost between nodes v_3 and v_4.

	v_0	v_1	v_2	v_3	v_4	v_5	v_6	v_7
v_0:	$(0,v_0)$	$(5,v_2)$	$(3,v_2)$	$(1,v_3)$	$(6,v_4)$	$(4,v_2)$	$(5,v_3)$	
v_1:	$(5,v_2)$	$(0,v_1)$	$(2,v_2)$	$(7,v_3)$		$(3,v_2)$	$(11,v_3)$	
v_2:	$(3,v_0)$	$(2,v_1)$	$(0,v_2)$	$(4,v_0)$	$(6,v_5)$	$(1,v_5)$		$(5,v_5)$
v_3:	$(1,v_0)$	$(7,v_1)$	$(4,v_0)$	$(0,v_3)$	$(7,v_0)$		$(4,v_6)$	$(6,v_6)$
v_4:	$(6,v_0)$		$(6,v_5)$	$(7,v_6)$	$(0,v_4)$	$(5,v_5)$	$(3,v_6)$	$(5,v_6)$
v_5:	$(4,v_2)$	$(3,v_2)$	$(1,v_2)$		$(5,v_4)$	$(0,v_5)$	$(6,v_7)$	$(4,v_7)$
v_6:	$(5,v_3)$	$(11,v_3)$		$(4,v_3)$	$(3,v_4)$	$(6,v_7)$	$(0,v_6)$	$(2,v_7)$
v_7:			$(5,v_5)$	$(6,v_6)$	$(5,v_6)$	$(4,v_5)$	$(2,v_6)$	$(0,v_7)$

Figure 5.10: The routing tables for the nodes in the undirected version of Figure 5.7, after two rounds of the Bellman-Ford algorithm.

Reconsider the routing table for node v_1. Again, v_2 will eventually pass its now updated table to v_1, reporting a cost of $d(v_2, v_3) = 4$ of a path it discovered to v_3. As soon as v_1 obtains this information, it will have found a better path to v_7 than the direct connection through edge $\langle v_1, v_3 \rangle$, namely via v_2. The cost for this path are $w(v_1, v_2) + d(v_2, v_3) = 2 + 4 = 6$. Hence, v_1 will adjust its routing table accordingly. Note that the only thing v_1 knows, is that messages for destination v_3 should be routed via v_2. In particular, v_1 is unaware of the *length* of its newly discovered path to v_3, i.e., the number of edges of that path.

The Bellman-Ford algorithm is particularly attractive because it allows each node to gradually discover optimal paths to the currently reachable nodes in the network. It is important to realize that the algorithm is completely **decentralized**: all decisions that a node takes concerning optimal routes is based entirely on local information, without the need to be *complete*. In contrast, Dijkstra's algorithm requires that the complete topology of the network is first disseminated to each node before each can start computing optimal routes (i.e., sink trees). Nevertheless, the algorithm had some serious drawbacks in practice, eventually making it less popular. Further information on applying the protocol in practice (where it is generally referred to as a **distance-vector routing protocol**) can be found in [Malkin and Steenstrup, 1995].

Note 5.6 (More information)
There is one particularly nasty problem inherent to the Bellman-Ford protocol. Consider a network in which the nodes are organized as a straight line:

Assume that the distance between two adjacent nodes is always 1 (i.e., $d(v_i, v_{i+1}) = 1$). Eventually, the nodes will build the following routing tables:

	Destination					
	v_1	v_2	v_3	v_4	v_5	v_6
$v_1:$	$(0, v_1)$	$(1, v_1)$	$(2, v_1)$	$(3, v_1)$	$(4, v_1)$	$(5, v_1)$
$v_2:$	$(1, v_1)$	$(0, v_2)$	$(1, v_3)$	$(2, v_3)$	$(3, v_3)$	$(4, v_3)$
$v_3:$	$(2, v_2)$	$(1, v_2)$	$(0, v_3)$	$(1, v_4)$	$(2, v_4)$	$(3, v_4)$
$v_4:$	$(3, v_3)$	$(2, v_3)$	$(1, v_3)$	$(0, v_4)$	$(1, v_5)$	$(2, v_5)$
$v_5:$	$(4, v_4)$	$(3, v_4)$	$(2, v_4)$	$(1, v_4)$	$(0, v_5)$	$(1, v_6)$
$v_6:$	$(5, v_5)$	$(4, v_5)$	$(3, v_5)$	$(2, v_5)$	$(1, v_5)$	$(0, v_6)$

Now suppose that the link between node v_1 and v_2 breaks. In other words, v_2 can no longer directly reach v_1. As a consequence, node v_2 will have to discover an alternative route to v_1, and "fortunately," notices that v_3 is advertising that it has a path to v_1 of routing cost 2. Of course, this advertised path is $[v_3, v_2, v_1]$, but this information is withheld from v_2. The only thing that v_2 gets to know from v_3 is that the latter has discovered a path to v_1 of cost 2. Node v_2 will then update its routing-table entry for getting to node v_1 with $(3, v_2)$ and advertise that it has discovered a path to v_1 of cost 3.

The problem will now become clear: v_3 had registered the entry $(2, v_2)$ based on the *initially* advertised routing cost by v_2 (which was 1), and its own routing cost of getting to v_2 (also 1). Now that v_2 is advertising a routing cost of 3, v_3 will adjust its entry to $(4, v_2)$, and subsequently advertise a routing cost of 4 to get to v_1. As soon as this new routing-cost information reaches v_2, it will adjust its advertised cost from 3 to 5. This process will not stop as long as the link between v_1 and v_2 remains defect. The result is known as the **count-to-infinity problem** which turned out to have no easy fix. In practical settings, the Bellman-Ford algorithm is used with a full advertisement of the path, allowing a node to discover whether it is part of that path, avoiding the mistake of choosing a path with a known broken link.

A note on algorithmic performance

Realizing that Dijkstra's algorithm as well as the Bellman-Ford algorithm lie at the heart of some of the most important routing algorithms in the Internet, it is worthwhile seeing how *efficient* these algorithms actually are. In particular, we can ask ourselves how long it will take to find a sink tree as a function of the number of vertices. As it turns out, in most cases Dijkstra's algorithm will outperform the Bellman-Ford solution. In particular, when graphs are large and have many edges, Dijkstra will generally be more efficient. To illustrate, Figure 5.11 shows the time to compute a sink tree as a function of the size of the graph (expressed in the number of vertices). Figure 5.11(a) shows the results for a so-called **grid graph**: a graph in which the vertices and edges are organized as in a two-dimensional grid. Figure 5.11(b) shows the time needed to compute a sink tree in a complete graph. Indeed, we can see that Bellman-Ford outperforms Dijkstra's algorithm for grid graphs, but not for complete graphs.

These results are not so surprising when taking a closer look at the number of algorithmic steps that we need to take for each algorithm. Let us first consider Dijkstra's algorithm. At each step t, we need to inspect all vertices in $R_t(u) \setminus S_t(u)$, after which we expand $S_t(u)$ with one vertex. If n denotes the total set of vertices, then each step thus requires considering *in the order* of n vertices, which we repeat n times. In other words, we can expect that

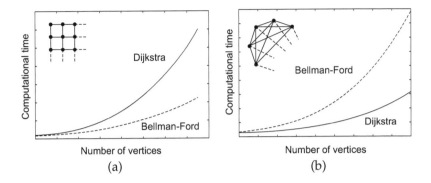

Figure 5.11: The time needed to compute a sink tree in (a) a grid graph and (b) a complete graph.

the computational time of Dijkstra's algorithm is roughly proportional to n^2.

For the Bellman-Ford algorithm, we observe something different. At each step, each vertex needs to inspect the information collected at each of its neighbors. In total, the vertices needs to inspect roughly m other vertices, where m is the total number of edges. The total number of steps we need to perform is equal to the length of the longest shortest path and can be shown to increase proportional to the number of vertices. Hence, the computational time of the Bellman-Ford algorithm is approximately proportional to $n \cdot m$.

We stress that these are merely *back-of-the-envelope* calculations. Indeed, when considering that the minimal number of edges that we need for a graph of n vertices to be connected is equal to $n - 1$ (as we showed in Theorem 5.1), we could equally argue that the Bellman-Ford algorithm will take at least also in the order of $n \cdot n - 1 \approx n^2$ time units to complete. More details need to be considered to arrive at more accurate calculations, but which goes beyond the scope of this text. What our calculations do show, is that the more edges a graph has, we may indeed expect that the Bellman-Ford algorithm performs comparatively less than Dijkstra's algorithm.

> **Note 5.7** (Mathematical language)
> Above, we stated that we needed to consider *in the order* of n vertices. This can be made mathematically precise using what is known as the **big O notation**, which allows us to describe the behavior of a function $f(x)$ for large values of x. The basic idea is that we want to capture what can be called the dominating components of a function. For example, the function $f(x) = ax^2 + bx + c$ is

5.4. ROUTING IN COMMUNICATION NETWORKS

completely dominated by the term ax^2 when x becomes very large. The other terms eventually hardly play a role anymore, regardless how big the constants b and c are. In fact, the *form* of $f(x)$ is completely determined by the term x^2.

Formally, we write $f(x) \sim \mathcal{O}(g(x))$ when there exists a constant $M > 0$ such that for all $x > x_0$ we have that $|f(x)| < M \cdot |g(x)|$. In other words, apart from a constant factor M, function $f(x)$ will always be bounded by function $g(x)$ after some value x_0 as shown in Figure 5.12.

We can also provide a lower bound for a function $f(x)$, in which case we write $f(x) \sim \Omega(g(x))$ meaning that there exists a constant M' such that for some value x'_0 we know that $|f(x)| > M' \cdot |g(x)|$. Note that $f(x) \sim \Omega(g(x))$ if and only if $g(x) \sim \mathcal{O}(f(x))$. Finally, a function $f(x)$ can eventually have exactly the same form as another function $g(x)$, or more precisely, there exist constants M and M' such that for all $x > x_0$ we have that $M' \cdot |g(x)| < |f(x)| < M \cdot |g(x)|$. In this case, we write $f(x) \sim \Theta(g(x))$. More information on computational complexity can be found in [Goodrich and Tamassia, 2002].

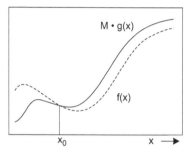

Figure 5.12: Bounding the function $f(x)$ by $g(x)$.

There is no doubt that the Bellman-Ford algorithm is elegant: it is fully decentralized, nodes need only publish their routing tables when an update occurs, and in practice it is generally just as efficient as Dijkstra's algorithm. Nevertheless, the algorithm is often less popular than Dijkstra's. One of its major problems is that when edge weights change often, nodes need to continuously adjust their routing tables, *and* propagate those changes throughout the network. If that propagation takes longer than the time between changes, we obviously have a problem. In the case of the Bellman-Ford algorithm, these problems can become so serious that constructing optimal sink trees is no longer possible, and special measures are needed. As it turns out, Dijkstra's algorithm is less susceptible to these propagation issues. Both type of algorithms continue to play a key role in the design and implementation of communication networks.

As a side note, both algorithms are considered to be **computationally**

efficient, meaning that their running time is in the order of some polynomial function such as, n^2 or n^3. In contrast, problems that require algorithms with a computational effort that grows exponentially in the problem size (which in our examples is expressed in terms of the size of a graph), are called **computationally inefficient**. Unfortunately, many graph problems fall into this class, such as finding Hamilton cycles, or determining whether two graphs are isomorphic. A standard text that discusses these issues is [Garey and Johnson, 1979]

Chapter 6

Network analysis

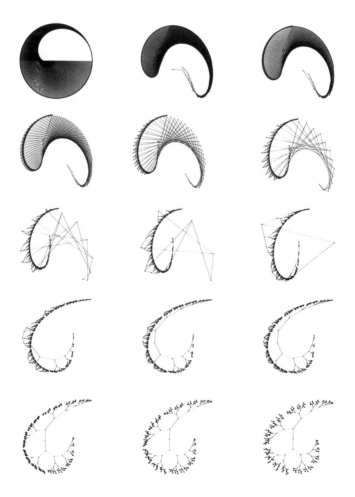

6.1. VERTEX DEGREES

Up to this point we have discussed some of the more elementary issues concerning graphs. In the real world, we are often confronted with a network and wish to examine some of its properties in order to get more insight in what we're actually dealing with. This is particularly true when dealing with large networks that exhibit apparently random structures. In the following chapters we will have a closer look at many of these networks, but before doing so, we take a look at some of the basic techniques that we can use to analyze those structures.

Network analysis is an emerging field of research, often founded on the use of various mathematical tools and methods (see, e.g., Brandes and Erlebach [2005]), and is also considered as a subarea of what is known as data mining of graphs [Cook and Holder, 2007]. In the following, we consider several metrics used in a myriad of sciences to analyze networks. We start with focusing on vertex degrees, followed by taking a closer at so-called distance statistics. An important concept that is used to characterize many real-world networks is clustering, which is discussed next. After that, we pay attention to the notion of centrality, which is particularly important for social networks.

6.1 Vertex degrees

Perhaps one of the simplest starting points for network analysis is taking a look at vertex degrees. As we know from Theorem 2.4, the minimal vertex degree is an upper bound for the vertex and edge connectivity of a graph. However, there are other properties to examine through vertex degrees. For example, using degrees allows us to identify the key players in social networks: those nodes with a high vertex degree.

Also, degrees, and notably degree sequences can be used to derive information on the *structure* of a network. For example, if most vertex degrees are the same, we are dealing with a more or less regular network in which vertices have equal roles. On the other hand, with very skewed degree sequences, that is, sequences in which a few vertices have relatively high degrees in comparison to others, these high-degree vertices play the role of **hubs**, of which the removal may actually partition a connected network into several components.

Finally, as we already discussed, if we are to test isomorphism between two graphs, we can start with testing whether their respective degree sequences are the same. If they are not, then Theorem 2.3 tells us that they cannot be isomorphic. In the following, we first take a look at degree distributions, followed by a few words on degree correlations.

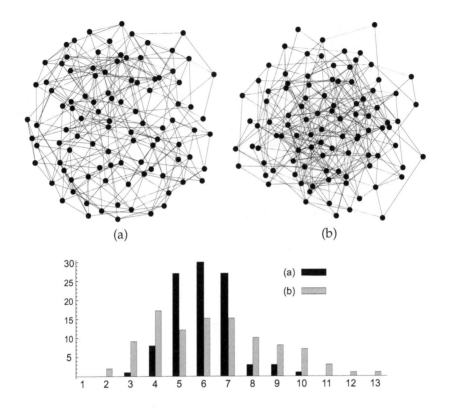

Figure 6.1: Two different graphs $GA_{complex}$ and $GB_{complex}$ and their respective histograms of vertex degrees.

Degree distribution

A degree sequence can often best be plotted by means of a histogram. In that case, for a simple, connected graph having n vertices, we plot the values $h(d) \stackrel{\text{def}}{=} |\{v \in V(G) | \delta(v) = d\}|$. In other words, $h(d)$ is the number of vertices having degree d. If for some value D we have that $h(d) = 0$ for all $d > D$, we simply discard those $h(d)$. Obviously, $\sum_{d=0}^{n-1} h(d) = n$. To illustrate, consider the graphs in Figure 6.1, which we will denote as $GA_{complex}$ and $GB_{complex}$, respectively. From the figure, we may suspect that they are different (and, in fact, if we consider other embeddings this difference will be more evident), but expressing this difference may be somewhat difficult. However, when considering their respective degree distributions, we see that we are indeed dealing with two very different graphs. To complete this simple analysis, we note that both graphs have 100 vertices, with the graph

6.1. VERTEX DEGREES

from Figure 6.1(a) having 300 edges, and the one from (b) having 317 edges.

There are different ways to visualize degree distributions. Above we used histograms. We can also consider the *fraction* of vertices that have a certain degree, i.e., draw $h(d)/n$. This technique is actually used to approximate the probability $\mathbb{P}[\delta(u) = d]$ that a vertex u has degree d. Another technique is to first order the vertices according to their degree, and then plot the degree vertex. Effectively, we consider the degree sequence $[d_1, d_2, \ldots, d_n]$ of a graph and subsequently plot d_k for each k. To illustrate, Figure 6.2 shows this alternative way of displaying vertex degrees for our two example graphs from Figure 6.1.

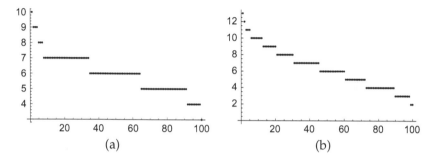

Figure 6.2: Visualizing the vertex degrees of $GA_{complex}$ and $GB_{complex}$ after ranking the vertices according to their degree. The y-axis shows the vertex degree, the x-axis the respective vertex rank.

When displaying vertex degrees, we sometimes also need to consider the scaling of the axis. Consider the following example of a 10,000-node graph, as shown in Figure 6.3 (which we discuss in more detail in Chapter 7). As in our previous example, we rank the vertices according to their degree and subsequently plot the vertex degree of each kth vertex. In Figure 6.3(a) we have used a linear scale for both axes. Unfortunately, we see that most vertices have the same, low degree, implying that it is difficult to see what is going on.

In Figure 6.3(b) we have used logarithmic scales for both axes. In other words, the *displayed* distance between two points on an axis is proportional to the logarithm of the *actual* distance between those two points. To illustrate, the displayed distance between $x = 10$ and $x = 100$ is the same as the one between $x = 100$ and $x = 1000$. The result is dramatic: we can now easily imagine that a straight line through all the data points can be drawn, implying that the vertex degree distribution follows some kind of exponential function. We will return to these issues in Chapter 7. In general, displaying the distribution of vertex degrees in many cases provides a

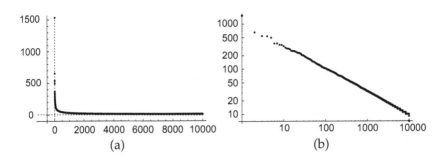

Figure 6.3: Different representations of visualizing vertex degrees: (a) using linear scales for the axes, and (b) using logarithmic scales.

lot of information, and we shall make use of this technique quite often.

> **Note 6.1** (More information)
> In many cases, being able to display a vertex degree distribution allows us to more adequately apply a technique known as **curve fitting**. This is a well-known statistical technique by which we try to find a (continuous) function $f(x)$ through a set of data points, such that the total error we make is minimal. To explain, consider the degree sequence $[d_1, d_2, \ldots, d_n]$. In this case, we have n data points (k, d_k). When finding a suitable curve through these data points, we will be generally looking for a relatively simple function $f(x)$, in turn implying that we will not always have an exact fit for every data point. In other words, for every value of k there will be a difference between $f(k)$ and d_k. In practice, we then try to find a function that minimizes the so-called least square error ϵ:
> $$\epsilon = \sum_{k=1}^{n} \left(d_k - f(d_k)\right)^2$$
> Other error metrics are also possible. Most packages for data analysis or data plotting have facilities on board for simple and often also advanced curve fitting. We will not delve into any further details. More information on the technicalities can be found in [Judd et al., 2009].

Degree correlations

Besides just displaying vertex degrees, we are often interested to what extent vertices of the same or different degrees are also joined. For example, in social networks high-degree vertices seem to generally be joined to each other, whereas in many technological networks, high-degree vertices are joined with low-degree ones [Newman, 2002]. The underlying phenomenon

6.1. VERTEX DEGREES

that we are dealing with is that in real-world networks we often see that similar nodes tend to link to each other, or, in contrast, that there is a tendency for dissimilar nodes to have links. The extent to which this phenomenon occurs is known as **assortative mixing**. Similarity is defined by all kinds of network-specific properties: the subject of Web pages, the preferences or taste of people, the number of shared files in peer-to-peer computer networks, etc. These properties are normally not captured when modeling real-world networks. At best, we can assign a **type** to a vertex and then ask ourselves to what extent vertices of the same or different type are joined (as is discussed by Newman [2003b]).

A much simpler approach is to consider only the vertex degree and to measure the **degree correlation** between the respective degrees of two adjacent vertices. Informally, the correlation between two variables x and y tells us to what extent we can expect that if we see a change in x, we will also see a change in y. If the correlation is positive, then an increase in x should show us also an increase in y. In the case of a negative correlation, an increase in x will show a decrease in the value of y. It is important to realize that we are dealing with *observed* changes. In other words, x and y are two observable variables such as humidity and the growth of a plant in the case of a biological system.

Formally, correlation is defined through what is known as a correlation coefficient:

Definition 6.1: *Let x and y be two stochastic variables, for which we have a series of observation pairs $(x_1, y_1), (x_2, y_2), \ldots, (x_n, y_n)$. The **correlation coefficient** $r(x, y)$ between x and y is defined as:*

$$r(x,y) \stackrel{\text{def}}{=} \frac{\frac{1}{n}\sum_{i=1}^{n}((x_i - \bar{x})(y_i - \bar{y}))}{\sqrt{\frac{1}{n}\sum_{i=1}^{n}(x_i - \bar{x})^2} \cdot \sqrt{\frac{1}{n}\sum_{i=1}^{n}(y_i - \bar{y})^2}}$$

where $\bar{x}$ is the average over the x_i's: $\bar{x} \stackrel{\text{def}}{=} \frac{1}{n}\sum_{i=1}^{n} x_i$, and likewise $\bar{y} \stackrel{\text{def}}{=} \frac{1}{n}\sum_{i=1}^{n} y_i$.

Note that the expression for $r(x, y)$ can be slightly simplified to

$$r(x,y) \stackrel{\text{def}}{=} \frac{\sum_{i=1}^{n}((x_i - \bar{x})(y_i - \bar{y}))}{\sqrt{\sum_{i=1}^{n}(x_i - \bar{x})^2} \cdot \sqrt{\sum_{i=1}^{n}(y_i - \bar{y})^2}}$$

Note 6.2 (Mathematical language)
If you have never seen formal definitions of correlations before, they can be quite intimidating. For our purposes, it is merely important that you have some intuition of where they come from. First, consider the expression

$\sum((x_i - \bar{x})(y_i - \bar{y}))$. Each term $(x_i - \bar{x})$ measures to what extent the observed value x_i deviates from the average observed values of x. If x and y are positively correlated, we would expect to see that each product $(x_i - \bar{x})(y_i - \bar{y})$ would also be positive (and certainly nonzero). In essence, the only thing that we are doing is simply computing the average over all these products, for which reason we divide the sum by the total number of observations, n.

So what are these terms in the denominator? As we just mentioned, $(x_i - \bar{x})$ measures the deviation of x_i from the average over all observations. In order to truly compare such deviations, we need to *normalize* our measurements. In other words, we need to make sure that the *ranges* of values that we are comparing are more or less the same, otherwise we will be biasing our measurements towards the variable with the largest ranges. One approach is to simply divide our observations by the average deviation, that is, $\frac{1}{n}\sum(x_i - \bar{x})$. However, for reasons that are beyond the scope of this text, it is common practice to use a different "average," namely $\sqrt{\frac{1}{n}\sum(x_i - \bar{x})^2}$, which is known as the **standard deviation**.

It should be noted that this explanation does not do just to the mathematical statistics underlying the definition of the correlation coefficient. In fact, the definition should actually be fine-tuned. More information can be found in Mandel [1984] or Judd et al. [2009].

Taking this formal definition of correlation as our basis, we can now define the correlation between vertex degrees. To this end, we make use of a graph's adjacency matrix $\mathbf{A}$. Recall that for a simple graph G with vertex set $V(G) = \{v_1, v_2, \ldots, v_n\}$, $\mathbf{A}[i,j] = 1$ if there is an edge joining vertex v_i and v_j, and otherwise $\mathbf{A}[i,j] = 0$.

Definition 6.2: *Let G be a simple graph with degree sequence $\mathbf{d} = [d_1, d_2, \ldots, d_n]$ and adjacency matrix $\mathbf{A}$. Let $V(G) = \{v_1, v_2, \ldots, v_n\}$ be such that $\delta(v_i) = d_i$. The **degree correlation** of G is defined as:*

$$r_{deg}(G) \stackrel{def}{=} \frac{\sum_{i=1}^{n}\sum_{j=i+1}^{n}\left((d_i - \bar{d})(d_j - \bar{d}) \cdot \mathbf{A}[i,j]\right)}{\sum_{i=1}^{n}(d_i - \bar{d})^2}$$

where $\bar{d}$ denotes the average vertex degree, i.e., $\frac{1}{n}\sum_{i=1}^{n} d_i$.

The similarity between $r(x,y)$ and $r_{deg}(G)$ should be obvious. Except for the use of the adjacency matrix, it is seen that the form of the respective nominators is virtually the same, with $\mathbf{d}$ essentially replacing both x and y in $r(x,y)$. That the same holds for the denominator can be seen when considering that

$$\sqrt{\frac{1}{n}\sum(d_i - \bar{d})^2} \cdot \sqrt{\frac{1}{n}\sum(d_i - \bar{d})^2} = \frac{1}{n}\sum(d_i - \bar{d})^2$$

6.1. VERTEX DEGREES

> **Note 6.3** (Mathematical language)
> Note how we used the adjacency matrix **A** to elegantly sum up all possible edges between two vertices, but discarding those that are not part of G. An equivalent, yet more concise notation is the following:
>
> $$r_{deg}(G) \overset{\text{def}}{=} \frac{\sum_{j>i}((d_i - \bar{d})(d_j - \bar{d}) \cdot \mathbf{A}[i,j])}{\sum_{i=1}^{n}(d_i - \bar{d})^2}$$
>
> in which case we assume that the exact values of i and j are clear from the context in which the summation is used.
>
> For an alternative notation in which the adjacency matrix is not used at all, we assume that the edges in G are indexed such that $e_{i,j} \in E(G)$ if and only if (1) there is an edge joining vertex v_i and v_j, and (2) $i > j$. This brings us to:
>
> $$r_{deg}(G) \overset{\text{def}}{=} \frac{\sum_{e_{i,j}}((d_i - \bar{d})(d_j - \bar{d}))}{\sum_{i=1}^{n}(d_i - \bar{d})^2}$$
>
> The drawback of this notation is that it is less explicit in exactly which vertex degrees we should take into account. On the other hand, you could argue that it expresses more concisely what degree correlation is.

An even simpler metric for capturing vertex correlations is proposed by Li et al. [2005] who define the scale-freeness of a graph:

Definition 6.3: *Let G be a simple graph with degree sequence* $[d_1, d_2, \ldots, d_n]$ *and adjacency matrix* **A**. *Let* $V(G) = \{v_1, v_2, \ldots, v_n\}$ *be such that* $\delta(v_i) = d_i$. *The* ***scale-freeness*** $s(G)$ *of G is defined as*

$$s(G) = \sum_{i=1}^{n} \sum_{j=i+1}^{n} (d_i \cdot d_j \cdot \mathbf{A}[i,j])$$

An important observation is that $s(G)$ is maximal when high-degree vertices are connected to each other. In other words, the scale-freeness is larger when hubs are attached to other hubs, forming a kind of cluster. However, the drawback of the form just given, is that it makes it difficult to compare graphs with each other. Therefore, we again need some kind of normalization. This can be achieved by considering what the maximal attainable scale-freeness is for all graphs with the same degree sequence:

Definition 6.4: *Let G be a simple graph with degree sequence* $\mathbf{d} = [d_1, d_2, \ldots, d_n]$ *and adjacency matrix* **A**. *Let* $V(G) = \{v_1, v_2, \ldots, v_n\}$ *be such that* $\delta(v_i) = d_i$. *Let* $\mathcal{G}(\mathbf{d})$ *be the collection of graphs with degree sequence* $\mathbf{d}$. *The* ***normalized***

scale-freeness $S(G)$ of G is defined as

$$S(G) = \frac{\sum_{i=1}^{n} \sum_{j=i+1}^{n} (d_i \cdot d_j)}{\max\{s(H)|H \in \mathcal{G}(\mathbf{d})\}}$$

Of course, the problem in this case is to find the maximal scale-freeness, which boils down to finding a graph H having degree sequence $\mathbf{d}$ and a maximal value $s(H)$. The procedure is too involved for our purposes, and the interested reader is referred to Li et al. [2005] for further information.

6.2 Distance statistics

Besides vertex-degree distributions, various distance statistics form an important class for network analysis. The distance between two vertices v and w in a graph is expressed in terms of the length of the shortest path between v and w.

Definition 6.5: *Let G be a directed or undirected graph and $u, v \in V(G)$. The **(geodesic) distance** between u and v, denoted as $d(u,v)$, is the length of a shortest (u,v)-path.*

Note that we have given an alternative definition for distance: in the case of weighted graphs, the distance between two vertices u and v is generally defined in terms of a (u,v)-path having minimal weight. The *length* of such a path, however, need not be minimal. In practice, which type of distance is meant is generally easy to understand from the context in which it is used. Furthermore, we discussed in Chapter 5 how to compute shortest paths, and demonstrated that there are efficient ways to find those paths. Note that in an undirected graph, $d(u,v) = d(v,u)$, but that this need not be the case for a directed graph.

What can we learn from distance statistics? Again, they can be used to see to what extent two networks are different or not, but also to give an indication of the relative importance of each of the nodes in a network. Let us first consider a few simple metrics (see also [Brinkmeier and Shank, 2005]). The **eccentricity** of a vertex u tells us how far the farthest vertex from u is positioned in the network. The **radius** of a network, defined as the minimum over all eccentricity values, is an indication of how disseparate the vertices in a network actually are. Finally, the **diameter** simply tells what the maximal distance in a network is. Formally, we have:

Definition 6.6: *Consider a connected graph G and let $d(u,v)$ denote the distance between vertices u and v. The **eccentricity** $\epsilon(u)$ of a vertex u in G is defined as $\max\{d(u,v)|v \in V(G)\}$. The **radius** $rad(G)$ is equal to $\min\{\epsilon(u)|u \in V(G)\}$.*

6.2. DISTANCE STATISTICS

*Finally, the **diameter** of G is the maximal shortest path between any two vertices:*
$$diam(G) = \max\{d(u,v) | u, v \in V(G)\}.$$

Note that these definitions apply to directed as well as undirected graphs.

Although the diameter gives us useful information, it may not be powerful enough to discriminate among graphs. An equally important and related metric for network analysis is to consider the distribution of path lengths. In particular, The *average* distance between vertices can provide useful information.

Definition 6.7: *Let G be a connected graph with vertex set V, and let $\bar{d}(u)$ denote the average length of the shortest paths from vertex u to any other vertex v in G:*

$$\bar{d}(u) \stackrel{\text{def}}{=} \frac{1}{|V|-1} \sum_{v \in V, v \neq u} d(u,v)$$

*The **average path length** $\bar{d}(G)$ is defined as*

$$\bar{d}(G) \stackrel{\text{def}}{=} \frac{1}{|V|} \sum_{u \in V} \bar{d}(u) = \frac{1}{|V|^2 - |V|} \sum_{u,v \in V, u \neq v} d(u,v)$$

*The **characteristic path length** of G is defined as the median over all $\bar{d}(u)$.*

Note 6.4 (Mathematical language)
Recall that the **median** over a set of n nondecreasing values $x_1, x_2, \ldots, x_n$ is equal to $x_{(n+1)/2}$ in case n is odd. If n is even, the median is often taken equal to $(x_{n/2} + x_{n/2+1})/2$. In other words, the median separates the higher values from the lower values into two equally-sized subsets. As we shall see later, the characteristic path length is particularly important when dealing with networks with only a few high-degree vertices and many low-degree vertices.

Note 6.5 (More information)
Why even bother about the characteristic path length? The problem with the average path length is that its computation becomes quite cumbersome for very large graphs. As we explained in Chapter 5, the time to compute all shortest paths to a given vertex following Dijkstra's algorithm is roughly proportional to n^2, with n being the number of vertices. In order to compute the average path length, we need to compute the shortest paths between *all* pairs of vertices, of which the computational effort is proportional to roughly n^3. It is not difficult to imagine that for large graphs, with, say more than a few thousand vertices, this can indeed be rather time-consuming. To illustrate, such a computation for a 10,000-node network can easily take tens of hours on a modern desktop computer.

As an alternative, we can also try to estimate the average path length. As it turns out, there are extremely efficient techniques to do this for the characteristic path length, but not for the average path length. Considering that for many cases the two metrics return approximately the same value, considering the characteristic path length is often preferred.

Let us consider these metrics for the graph G_{simple} shown in Figure 6.4. Regarding the eccentricity of each vertex and average distances between vertices, these can be easily derived by considering the length of the shortest paths between pairs of vertices, as shown in Figure 6.4. As a consequence, the radius of the graph is equal to 5, whereas the diameter is equal to 9. Likewise, we can compute the average path length of the graph to be 4.29. By ordering the average path lengths of the vertices, we obtain the sequence [3.17, 3.50, 3.67, 4.00, 4.33, 5.33, 6.00], from which we compute the characteristic path length to be 4.

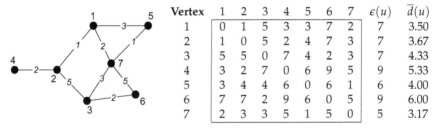

Figure 6.4: The distance between vertices of the graph G_{simple} (left) and the resulting eccentricity and average path lengths.

To complete this section, for our graphs from Figure 6.1 we find the following values for these distance metrics, again illustrating that we are indeed dealing with two very different graphs:

Metric	$GA_{complex}$	$GB_{complex}$
Average eccentricity	4.59	4.09
Radius	4	3
Diameter	6	5
Average path length	2.96	2.67
Characteristic path length	2.95	2.63

6.3 Clustering coefficient

Another, often used metric is what is known as the clustering coefficient. The idea behind this coefficient is rather simple: we want to see, for a given vertex v, to what extent the neighbors of v are also neighbors of each other. In other words, to what extent are vertices adjacent to v also adjacent to each other. Before we delve into all kinds of formalities, let us briefly consider why measuring clustering is important.

Some effects of clustering

A common way toward spreading information is simply having a node update its neighbors. In turn, neighbors can inform their neighbors, and so on. There are many variations to this model, such as having a node select only one or a few of its neighbors, or deciding to stop spreading updates when it notices that a selected neighbor already has the information. Informally, this type of dissemination is often described in the form of **gossiping models**, also known as **epidemic dissemination** [Eugster et al., 2004]. The model is very general: instead of information we can also consider spreading of diseases, but also viruses over the Internet. Another example is that of forming of opinions, which often depends on what the majority of your community thinks. We shall return to these issues in more detail when discussing peer-to-peer networks in Chapter 8.

When considering real-world networks, we often see that they are organized as a collection of interconnected groups. In terms of social networks, this means that we can often clearly distinguish communities of nodes with many links between its members, yet relatively few links between nodes that belong to different communities. Actually indicating which nodes belong to which communities may not be easy at all. Also, nodes generally belong to more than one community. However, we can express the *existence* of communities by means of a clustering coefficient. As shown by Xu and Liu [2008], it turns out that there is a clear relationship between the speed by which information is disseminated in social networks and the clustering coefficient: the higher the degree of clustering, the slower the dissemination. To a certain extent, this result may seem quite obvious, but from a formal (i.e., mathematical) point of view, it turns out to be not so trivial.

What this means is that if we want to design a dissemination protocol, we may need to take special measures in highly clustered networks in order to guarantee a certain performance regarding the dissemination speed. This alone has been enough reason for researchers to define and measure the clustering coefficient of a network.

Besides this reason, measuring the clustering coefficient obviously allows us to simply compare different networks, without necessarily wanting to make use of the actual values of the respective coefficients. In this sense, clustering coefficients can help in classifying networks.

Local view

We first consider clustering from the perspective of vertices, as originally introduced by Watts and Strogatz [1998]. From this so-called **local view**, the best clustering that we can achieve is that all neighbors are adjacent to each other. In other words, the neighbor set $N(v)$ of v forms a complete graph. Letting $n_v = |N(v)|$, we know that $N(v)$ will have a maximum of $\binom{n_v}{2} = \frac{1}{2}n_v(n_v - 1)$ edges. For the clustering coefficient, we then simply take a look at the ratio between the actual number of edges and the attainable maximum.

Definition 6.8: *Consider a simple connected, undirected graph G and vertex $v \in V(G)$ with neighbor set $N(v)$. Let $n_v = |N(v)|$ and m_v be the number of edges in the subgraph induced by $N(v)$, i.e., $m_v = |E(G[N(v)])|$. The **clustering coefficient** $cc(v)$ for vertex v with degree $\delta(v)$ is defined as*

$$cc(v) \stackrel{\text{def}}{=} \begin{cases} m_v / \binom{n_v}{2} = \frac{2 \cdot m_v}{n_v(n_v-1)} & \text{if } \delta(v) > 1 \\ \text{undefined} & \text{otherwise} \end{cases}$$

Note that we require that a vertex is adjacent to at least two other distinct vertices. Taking this into account, the clustering coefficient $CC(G)$ for the entire graph is defined as the average over all (well defined) clustering coefficients of its vertices:

Definition 6.9: *Consider a simple connected graph G. Let V^* denote the set of vertices $\{v \in V(G) | \delta(v) > 1\}$. The **clustering coefficient** $CC(G)$ for G is defined as*

$$CC(G) \stackrel{\text{def}}{=} \frac{1}{|V^*|} \sum_{v \in V^*} cc(v)$$

This notion of clustering can easily be extended to directed graphs, in which case we merely need to distinguish the case that we have an arc $\langle \overrightarrow{v,w} \rangle$ from v to w from an arc $\langle \overrightarrow{w,v} \rangle$. The neighbor set $N(v)$ of a vertex v will have a maximum number of $2 \cdot \binom{n_v}{2} = n_v(n_v - 1)$ arcs, i.e., twice as many arcs in comparison to the number of edges in the undirected case. This brings us to:

Definition 6.10: *Let D be a simple connected, directed graph D. Consider vertex $v \in V(D)$ with neighbor set $N(v)$. Let $n_v = |N(v)|$ and m_v be the number of*

6.3. CLUSTERING COEFFICIENT

arcs in the subgraph induced by $N(v)$, i.e., $m_v = |A(G[N(v)])|$. The **clustering coefficient** $cc(v)$ for vertex v with degree $\delta(v) = \delta_{in}(v) + \delta_{out}(v)$ is defined as

$$cc(v) \stackrel{\text{def}}{=} \begin{cases} m_v / \left(2 \cdot \binom{n_v}{2}\right) = \frac{m_v}{n_v(n_v-1)} & \text{if } \delta(v) > 1 \\ \text{undefined} & \text{otherwise} \end{cases}$$

In our definition for the clustering coefficient of a graph we did not make a distinction between directed and undirected graphs. Indeed, the definition stays the same.

Now consider the case of a weighted, undirected graph. As we mentioned, the clustering coefficient indicates the extent to which nodes in a network form (more or less) closed groups. If weights represent the intensity by which, for example, interactions take place, then weights are also indicative for the strength, or closedness of a group. This reasoning motivated Barrat et al. [2004] to introduce a weighted clustering coefficient. To this end, rather than merely considering the degree of a vertex v, they first take into account a weighted form of the vertex degree, called the vertex strength:

Definition 6.11: *Consider a simple weighted undirected graph G with vertex set $V(G) = \{v_1, v_2, \ldots, v_n\}$ and adjacency matrix $\mathbf{A}$. The **vertex strength** $\sigma(v_i)$ of vertex v_i is defined as the total sum of the weights of edges incident with v_i:*

$$\sigma(v_i) \stackrel{\text{def}}{=} \sum_{j=1}^{n} w(\langle v_i, v_j \rangle) \cdot \mathbf{A}[i,j]$$

We can now define the weighted clustering coefficient as follows.

Definition 6.12: *Consider a simple weighted undirected graph G with vertex set $V(G) = \{v_1, v_2, \ldots, v_n\}$ and adjacency matrix $\mathbf{A}$. The **weighted clustering coefficient** $cc(v_i)$ of vertex v_i is defined as:*

$$cc(v_i) \stackrel{\text{def}}{=} \begin{cases} \dfrac{\sum_{e_{i,j}, e_{i,k} \in E(G)} \left(w(e_{i,j}) + w(e_{i,k})\right) \cdot \mathbf{A}[i,j] \cdot \mathbf{A}[i,k] \cdot \mathbf{A}[j,k]}{2 \cdot \sigma(v_i) \left(\delta(v_i) - 1\right)} & \text{if } \delta(v_i) > 1 \\ \text{undefined} & \text{otherwise} \end{cases}$$

where $e_{i,j}$ is the edge joining v_i and v_j.

In other words, we consider only those edges $\langle u, v \rangle$, $\langle u, w \rangle$ incident with u, whose other end points, v and w, respectively, are joined as well. We leave it as an exercise to the reader to show that in the special case that all weights are equal to 1, the weighted clustering coefficient is equal to the clustering coefficient for an unweighted graph.

> **Note 6.6** (Mathematical language)
> The notations used for the last clustering coefficient may appear somewhat intricate. Let's inspect them a bit further. First note how we have again conveniently made use of the adjacency matrix to simplify our notation. In the expression
>
> $$\sigma(v_i) \stackrel{\text{def}}{=} \sum_{j=1}^{n} w(\langle v_i, v_j \rangle) \cdot \mathbf{A}[i,j]$$
>
> $\mathbf{A}[i,j]$ will be equal to 0 when there is no edge joining vertex v_i and v_j, effectively meaning that we will be ignoring the term $w(\langle v_i, v_j \rangle)$ (recall that for a nonexistent edge e, we let $w(e) = \infty$). An equivalent definition could have been formulated using the neighbor set $N(v)$ of vertex v, leading to:
>
> $$\sigma(v_i) \stackrel{\text{def}}{=} \sum_{v_j \in N(v_i)} w(\langle v_i, v_j \rangle)$$
>
> Somewhat more complicated is the actual expression for the clustering coefficient in a weighted undirected graph. In this case, the product $\mathbf{A}[i,j] \cdot \mathbf{A}[i,k] \cdot \mathbf{A}[j,k]$ in the nominator effectively allows us to consider only those cases in which vertices v_i, v_j, and v_k are all pairwise joined, i.e., forming a complete subgraph. For this **triangle**, we are actually interested in the edge joining v_i's neighbors v_j and v_k. The weight that we assign to the fact that these two neighbors are joined is determined entirely by how important v_j and v_k are to v_i, which is expressed by the respective weights of the edges $e_{i,j} = \langle v_i, v_j \rangle$ and $e_{j,k} = \langle v_j, v_k \rangle$. In the end, the importance of the adjacency of v_j and v_k for v_i is simply expressed as the weight $w(e_{i,j}) + w(e_{j,k})$.
>
> A few other observations may further help understand the definition of clustering coefficient for weighted graphs. Note that because of the unordered way we are summing over edges, we will actually be considering all *pairs* of edges incident with v_i twice, and thus also the triangles at v_i. This explains the factor 2 in the denominator. Finally, the division by the strength of v_i will now put a *relative* weight on the importance of two of v_i's neighbors being adjacent, allowing the clustering around different vertices to be compared to each other.

Global view

As explained by Newman [2003a] there is a reasonable alternative definition for the clustering coefficient based on the number of triples and triangles in a graph G, which are defined as follows:

Definition 6.13: *Consider a simple, undirected graph G and a vertex $v \in V(G)$. A **triangle** at v is a complete subgraph of G with exactly three vertices, including v. A **triple** at v is a subgraph of exactly three vertices and two edges, where v is incident with the two edges.*

6.3. CLUSTERING COEFFICIENT

We will use the notations $n_\wedge(v)$ to denote the number of triples at v, and $n_\triangle(v)$ the number of triangles at v. Likewise, we can consider the total number $n_\triangle(G)$ of distinct triangles of a graph G and its number $n_\wedge(G)$ of distinct triples. We define the transitivity of a graph as follows:

Definition 6.14: *Let G be a simple, connected graph with $n_\triangle(G)$ distinct triangles and $n_\wedge(G)$ distinct triples. The **network transitivity** $\tau(G)$ is defined as the ratio $n_\triangle(G)/n_\wedge(G)$.*

Network transitivity is considered a **global view** on clustering, as it considers the network as a whole instead of the situation local to vertices. To illustrate these two approaches, let us return to graph G_{simple} from Figure 6.4. It is not difficult to see that for each vertex we have the following:

Vertex:	1	2	3	4	5	6	7
cc:	1/3	0	1/3	undefined	1	1	1/3
$n_\wedge$:	3	3	3	0	1	1	6

This leads to a clustering coefficient of $CC(G_{simple}) = 3/6$ for the graph itself. Regarding the transitivity, we need to first count the number of triangles, of which there are only two. The total number of distinct triples is 17 (by simply summing up $n_\wedge(v)$), which means that $\tau(G_{simple}) = 2/17$.

This method can, of course, also be applied to our larger examples from Figure 6.1, for which we find:

Metric	$GA_{complex}$	$GB_{complex}$
Clustering coefficient	0.209	0.049
Transitivity	0.064	0.019

The difference between clustering coefficient and network transitivity is subtle, yet important to make, if only for the reason that different communities often loosely speak about the clustering coefficient of a graph G without making clear whether they mean $CC(G)$ or $\tau(G)$. In the case of social networks, the clustering coefficient of a graph is also known as the network density, which is formally defined as follows [Hage and Harary, 1983; Wasserman and Faust, 1994]:

Definition 6.15: *Consider a simple, undirected graph G with n vertices and m edges. The **network density** $\rho(G)$ of G is defined as $m/\binom{n}{2}$.*

In other words, the network density tells us to what extent a graph is complete or not, which is intuitively what we also used for defining the clustering coefficient. However, it is fairly easy to see that the network density and clustering coefficient are not the same, which we leave as an exercise to the reader.

Note 6.7 (More information)
The two notions of clustering are clearly related, especially when considering that we can also define the clustering coefficient of a vertex in terms of triangles and triples. Clearly, we have

$$cc(v) = \frac{n_\Delta(v)}{n_\Lambda(v)} \quad \text{and also} \quad n_\Lambda(v) = \binom{\delta(v)}{2}$$

Furthermore, it should also be clear that

$$n_\Delta(G) = \frac{1}{3} \sum_{v \in V^*} n_\Delta(v)$$

to account for the fact that each triangle is counted three times if we consider each vertex of the graph. However, only in special cases will we see that

$$\tau(G) = \frac{n_\Delta(G)}{n_\Lambda(G)} = \frac{\sum n_\Delta(v)}{3 \sum n_\Lambda(v)} \quad \text{and} \quad CC(G) = \frac{1}{|V^*|} \sum \left(\frac{n_\Delta(v)}{n_\Lambda(v)} \right)$$

are equal.

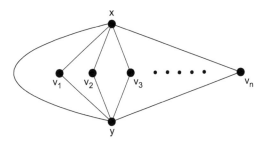

Figure 6.5: A graph with different clustering coefficient and transitivity.

The difference between the two metrics is also illustrated in Figure 6.5. Let G_k be the subgraph induced by vertices $\{x, y, v_1, v_2, \ldots, v_k\}$. It is not difficult to see that for every subgraph G_k we have

$$cc(u) = \begin{cases} 1 & \text{if } u = v_i \text{ for } 1 \leq i \leq k \\ \frac{k}{0.5 \cdot k(k+1)} = \frac{2}{k+1} & \text{if } u = x \text{ or } u = y \end{cases}$$

As a consequence, we see that

$$CC(G) = \frac{1}{k+2} \left(2 \cdot \frac{2}{k+1} + k \cdot 1 \right) = \frac{k^2 + k + 4}{k^2 + 3k + 2}$$

and thus

$$\lim_{k \to \infty} CC(G) = 1$$

6.3. CLUSTERING COEFFICIENT

> To compute network transitivity, we need to count the number of triangles, which is equal to k. With $n_\Lambda(v_i) = 1$ and $n_\Lambda(x) = n_\Lambda(y) = \binom{k+1}{2} = \frac{1}{2}k(k+1)$, we find that
> $$\tau(G) = \frac{k}{2 \cdot 0.5 \cdot k(k+1) + k} = \frac{1}{k+2}$$
> and thus
> $$\lim_{k \to \infty} \tau(G) = 0$$

We can extend the notion of transitivity to weighted graphs following an approach suggested by Opsahl and Panzarasa [2009]. In this case, we need to assign a weight to triples and triangles, after which we compute the transitivity of a graph by considering the ratio of the cumulative weights on the triangles and that of the triples. Let us start with defining precisely what the weight of a triple or triangle is.

Definition 6.16: *Let G be a simple, undirected weighted graph and consider vertex $v \in V(G)$. If H is a triple or a triangle at v where edges e_1 and e_2 are incident with v, then the **triple weight** $w_\Lambda(H)$ and **triangle weight** $w_\triangle(H)$, respectively is equal to the average of the weights of e_1 and e_2, i.e.,*

$$w_\Lambda(H) \stackrel{\text{def}}{=} \frac{1}{2}(w(e_1) + w(e_2)) \quad \text{and} \quad w_\triangle(H) \stackrel{\text{def}}{=} \frac{1}{2}(w(e_1) + w(e_2))$$

In principle, the triple of triangle weight can also be defined as, for example, $\max\{w(e_1), w(e_2)\}$, but we shall not consider such details here. Using these definitions, we can then define the transitivity of a weighted graph as follows.

Definition 6.17: *Let G be a simple, undirected weighted graph with $\mathbf{H}_\triangle$ its set of triangles, and $\mathbf{H}_\Lambda$ its set of triples. The **network transitivity** $\tau(G)$ is defined as*

$$\tau(G) \stackrel{\text{def}}{=} \frac{\sum_{H \in \mathbf{H}_\triangle} w_\triangle(H)}{\sum_{H \in \mathbf{H}_\Lambda} w_\Lambda(H)}$$

Note that this definition is identical to that of transitivity in an unweighted graph when setting weights equal to 1.

Finally, Opsahl and Panzarasa [2009] extend their definition of transitivity to directed graphs, be they weighted or not. In this case, they simply use the same definition of weights for triples and triangles, respectively, but restrict the enumeration of these subgraphs to so-called nonvacuous triples and transitive triangles:

Definition 6.18: *Consider a (strict) directed graph D. Let H be a triple at v, with its neighbors u and w in H. H is a **nonvacuous triple** if either $\langle \vec{u,v} \rangle, \langle \vec{v,w} \rangle \in A(H)$ or $\langle \vec{w,v} \rangle, \langle \vec{v,u} \rangle \in A(H)$. If H was a triangle at v, then H is **transitive** if $A(H) = \{\langle \vec{u,v} \rangle, \langle \vec{v,w} \rangle, \langle \vec{u,w} \rangle\}$ or $A(H) = \{\langle \vec{w,v} \rangle, \langle \vec{v,u} \rangle, \langle \vec{w,u} \rangle\}$.*

In other words, H as a triple is nonvacuous if there exists either a (u,w)-path via v or a (w,u)-path via v, and H as a triangle is transitive if w can be reached from u both through an arc $\langle \vec{u,w} \rangle$ and a path in H via v, or u can be reached from w through an arc $\langle \vec{w,u} \rangle$ and a directed path through v. Figure 6.6 shows all possible (non)vacuous triples and (non)transitive triangles.

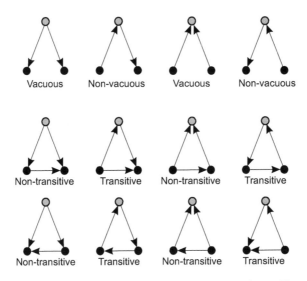

Figure 6.6: (Non)vacuous triples and (non)transitive triangles at (the marked) vertex v.

We will often use the clustering coefficient or network transitivity to compare different random graphs. Both metrics are used in practice, yet computing network transitivity for large graphs can be somewhat inefficient provided special measures are taken. We will not go into details here, but will return to various examples when discussing concrete examples of random graphs throughout the remaining chapters.

6.4 Centrality

Another important metric for network analysis is deciding on whether there are any vertices "more important" than others. The importance of a vertex

6.4. CENTRALITY

is, of course, dependent on what a graph is actually modeling. For example, when dealing with networks representing relationships between people, a vertex with a high degree may characterize an influential person. In a communication network, however, the importance of a vertex may be determined by the number of shortest paths of which it is member, for in that case it may be an indication of its workload regarding processing and forwarding messages.

In network analysis, this concept of importance is referred to as **centrality** [Kotschutzki et al., 2005]. Perhaps one of the simplest notions of centrality is identifying the center of a graph. It is formed by those vertices whose eccentricity is equal to the radius of a graph:

Definition 6.19: *Consider a (strongly) connected graph G. The **center** $C(G)$ of a graph G is the set of vertices with minimal eccentricity, i.e., $C(G) \stackrel{\text{def}}{=} \{v \in V(G) | \epsilon(v) = rad(G)\}$.*

Intuitively, a vertex is at the center of a graph when it is at minimal distance from all other vertices. Using the eccentricity of a vertex u, we can then define its centrality as:

Definition 6.20: *Let G be a (strongly) connected graph. The **(eccentricity based) vertex centrality** $c_E(u)$ of a vertex $u \in V(G)$ is defined as $1/\epsilon(u)$.*

All vertices in the center of a graph have maximal centrality, whereas indeed all vertices at the "edges" of a graph have very low centrality. Returning to graph G_{simple} from Figure 6.4, we find that the center consists only of vertex 7. With some computational effort, it can be shown that graph $GA_{complex}$ from Figure 6.1 has no less than 43 vertices in its center, whereas $GB_{complex}$ has only two vertices in the center.

Eccentricity can be used for determining whether certain functions in a network have been optimally placed. For example, when deciding on placing certain buildings in a city, we may want to take into account that those buildings should be conveniently reached, such as fire stations. In effect, the decision is to place certain functionality within a specific range of all nodes.

Eccentricity measures the maximum distance from one node to any other node in a network. In some cases, it is more important to know how close a node is to *all* other nodes. This means that we need to take into account all the distances from one node to the others. In that case, we simply take the total distance of that node to every node into account, as follows:

Definition 6.21: *Consider a (strongly) connected graph G. The **closeness** $c_C(u)$ of a vertex $u \in V(G)$ is defined as $c_C(u) \stackrel{\text{def}}{=} 1/\left(\sum_{v \in V(G)} d(u,v)\right)$.*

Returning to our example, it is clear that a fire station should be close to any arbitrarily chosen node. In that case, we want to optimize on the traveling distance when a fire breaks out. However, matters become different in the case of services that need to be accessed simultaneously from different nodes, such as with hospitals, a town hall, shopping centers, and so forth. This is where closeness comes into play. In those cases, we want to place a service conveniently close to as many nodes as possible, which is clearly a different criterion than minimizing the maximum distance that needs to be traveled.

For G_{simple} we find the following values for the closeness of its vertices. Although vertex 7 forms the center of G_{simple}, it is not the vertex closest to all others, which is vertex 1.

Vertex:	1	2	3	4	5	6	7
$\sum d(u,\cdot)$	21	22	27	32	24	37	29
$c_C(u)$:	0.048	0.045	0.037	0.031	0.042	0.027	0.034

Note that comparing closeness between vertices of different graphs may not be very useful. For example, when considering unweighted graphs, we see that the closeness of a vertex decreases as the graph consists of more vertices. For this reason, comparing the closeness of vertices is useful only relative to a given graph.

Vertex centrality and closeness are both related to the reachability of a vertex, and as such may indeed indicate the importance of a vertex. However, we have also seen another type of important vertices, namely cut vertices, whose removal actually partitions a graph. One can argue that such vertices form the center of a graph. Based on this observation, notably researchers in the social sciences have introduced what is referred to as **betweenness**. The basic idea is simple: if a vertex lies on many shortest paths connecting two other vertices, it is an important vertex. The reasoning is that the removal of such a vertex will directly influence the cost of the connectivity between other vertices, as other (i.e., longer) shortest paths will have to be followed. Formally, we have:

Definition 6.22: *Let G be a simple, (strongly) connected graph. Let $S(x,y)$ be the set of shortest paths between two vertices $x, y \in V(G)$, and $S(x, u, y) \subseteq S(x, y)$ the ones that pass through vertex $u \in V(G)$. The **betweenness centrality** $c_B(u)$ of vertex u is defined as*

$$c_B(u) = \sum_{x \neq y} \frac{|S(x,u,y)|}{|S(x,y)|}$$

Note that because G is (strongly) connected, $|S(x,y)| > 0$ for all pairs of distinct vertices x and y.

6.4. CENTRALITY

In the following chapters we will apply these and other metrics to specific types of graphs. As we'll see, more metrics can be defined to differentiate and characterize graphs, but many of these metrics are more easily explained and motivated given the specific context in which graphs and networks are used to model real-world situations.

CHAPTER 7

RANDOM NETWORKS

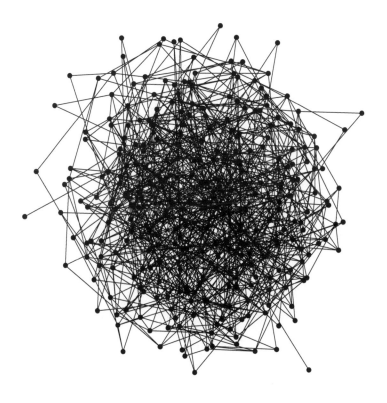

Up to this point we have largely covered the core of traditional graph theory. This core contains material that is mainly related to well-structured graphs, often of limited size, which is used for the type of applications we have discussed in the previous chapters. We now draw our attention to another type of graph for which the theoretical foundations were laid down in the late 1950s by Paul Erdös and Alfréd Rényi, namely graphs that were constructed by randomly adding edges. The field remained somewhat esoteric until the turn of the century when scientists began to discover that many natural phenomena could be described in terms of random graphs. This eventually lead to a boost in research on what have been coined *complex networks*, research that is found in a myriad of fields, ranging from neurology to traffic management to communication networks. Not without reason, this research is often referred to as *the new science of networks* [Barabási, 2002; Buchanan, 2002; Watts, 2004; Lewis, 2009].

In this chapter, we will take a first look at these random graphs (or random networks as they are more often called). It is also here that this book starts deviating from more traditional texts on graph theory.

7.1 Introduction

Intuitively, a random network is a (simple, connected) graph G in which pairs of vertices are connected by some probability. In general, this means that we start with a collection of n vertices and for each of the $\binom{n}{2}$ possible edges, we add edge $\langle u, v \rangle$ with some probability p_{uv}. In the simplest case, p_{uv} is the same for every pair of distinct vertices u and v.

Initially driven by curiosity only, random networks are now considered to be important for the simple reason that they allow us to model many real-world phenomena:

Spatial systems: In many cases, real-world networks have a spatial dimension in the sense that there is some notion of distance between nodes. Examples include railway networks, airline networks, computer networks, electricity networks, and neural networks. Modeling such networks as graphs implies that we need to let the probability of adding an edge be dependent on the distance between nodes in the real network: the larger the distance between two nodes, the smaller the probability of attaching them in the corresponding random graph. As it turns out, if we take this spatial dimension into account, along with some other properties that we discuss later on, random graphs can be used to accurately model real-world spatial networks.

Food webs: A food web (also called a food chain) describes the feeding relationships between organisms, that is, who eats whom. Obviously, we can model food webs as directed graphs. In particular, it turns out that in order to get insight in the resilience of ecosystems in terms of the extinction of species, modeling food webs as random graphs is an appropriate technique. Unlike many other real-world networks, food webs are generally relatively small (in the order of tens of a few hundreds of nodes). Also, there is controversy regarding their structure, which deviates from some of the more well-known random graphs (see, e.g., Dunne et al. [2002]). In this case, using techniques from network analysis as introduced in Chapter 2 and the theory of random graphs allows us to better understand the nature of food webs.

Collaboration networks: An important class of networks is formed by various collaborations between human beings. Famous is the analysis of networks of movie actors, formed by creating a graph of actors, linking any two who have ever played in the same movie (see also [Watts, 1999]). Likewise, it turned out that modeling collaborations between scientists using network analysis techniques and random-graph theory has provided insight in how science is formed. In particular, there is a body of work on citation networks, reflecting which scientific articles are cited in other articles. Such networks provide insight in the influence of published work.

In order to study real-world networks, it is necessary that we delve into the properties of random networks. These properties can be explained using the terminology that has been introduced so far, and form the basis for proper network analysis.

7.2 Classical random networks

As mentioned, random networks have been introduced and studied for several decades. Paul Erdös and Alfréd Rényi introduced what are now known as "classical" random networks, or **Erdös-Rényi networks** [Erdös and Rényi, 1959]. The basic idea is that we consider a simple, connected graph on n vertices, and that every two vertices are adjacent with some probability p. Erdös and Rényi introduced two different types of random graphs.

Definition 7.1: *An Erdös-Rényi model of a random network on n vertices, also referred to as an **ER random graph**, is an undirected graph $G_{n,p}$ in which each two (distinct) vertices are connected by an edge with probability p. For a given number*

7.2. CLASSICAL RANDOM NETWORKS

M of edges, the ER random graph $G_{n,M}$ is an undirected graph in which each of the M edges is incident to randomly chosen pairs of vertices.

It is important to note that two graphs $G_{n,p}^1$ and $G_{n,p}^2$ may be very different. Although they will both have n vertices, because an edge $e = \langle u, v \rangle$ between two vertices u and v exists only with a probability p, it may well be that $e \in E(G_{n,p}^1)$, yet that $e \notin E(G_{n,p}^2)$. In this light, we use the notation $ER(n,p)$ to denote the set of *all* ER random graphs with n vertices and probability p that two distinct vertices are joined.

Note that an $ER(n,p)$ graph is simple: there are no loops and there is at most one edge between two distinct vertices. In contrast, the formal definition of a $G_{n,M}$ random graph allows loops and multiple edges, but in practice we often see that they are restricted to their simple counterparts. In this book, we concentrate exclusively on $ER(n,p)$ graphs.

Degree distribution

Let us first see what we can expect when considering vertex degrees. For each vertex u of an $ER(n,p)$ graph, we know that there are at most $n-1$ other vertices to which it can be connected. Let $\mathbb{P}[\delta(u) = k]$ denote the probability that the degree of vertex u is k. Because there are a maximum of $n-1$ other vertices that can be a neighbor of u, it should be clear that there are $\binom{n-1}{k}$ possibilities for choosing k different vertices to be adjacent to u. The probability of having u joined with *exactly* k other vertices (and thus not with exactly $n-1-k$ vertices) is equal to $p^k \cdot (1-p)^{n-1-k}$, so that

$$\mathbb{P}[\delta(u) = k] = \binom{n-1}{k} p^k (1-p)^{n-1-k}$$

Note that our reasoning for the degree distribution of u applies to all vertices of an $ER(n,p)$ graph. Formally, this means that we can treat the vertex degree as a **random variable** δ, for which we have just shown that it follows what is known as a **binomial distribution**. In line with this observation, we can speak of the probability that a vertex degree has value k, and write $\mathbb{P}[\delta = k]$.

Note 7.1 (Mathematical language)
Probability and stochastics play an important role in random-graph theory, although we shall consider only a few concepts. The notion of a **random variable** is important. Intuitively, it is a variable whose values can each occur with a certain probability. In the case of **discrete random variables**, there are only a finite number of possible values. This is the case, for example, when consider-

ing the possible vertex degrees in an $ER(n,p)$ graph. Throughout this book, we consider only discrete random variables.

To characterize a (discrete) random variable X, we need to consider all its possible values. A simple example is where X denotes the possible outcomes of flipping a coin, for which there are only two possible outcomes: head or tail. Normally, each of these values can occur with equal probability, which is expressed as:

$$\mathbb{P}[X = \text{head}] = \mathbb{P}[X = \text{tail}] = \frac{1}{2}$$

Likewise, we can treat the vertex degree of an $ER(n,p)$ graph as a random variable δ with possible outcomes any value from the set $\{0, 1, \ldots, n-1\}$, and subsequently compute the probabilities $\mathbb{P}[\delta = k]$ for $0 \leq k < n$. In general, if we know that X can take values only from $\{x_1, x_2, \ldots, x_N\}$, it should be clear that

$$\sum_{i=1}^{N} \mathbb{P}[X = x_i] = 1$$

Given, for example, the distribution of the vertex degrees, we will regularly ask ourselves what the *average* vertex degree will be. For any discrete random variable X the notion of average is more accurately expressed in terms of its **mean**, also known as its **expected value**, defined as:

$$\mathbb{E}[X] \stackrel{\text{def}}{=} \sum_{i=1}^{N} x_i \cdot \mathbb{P}[X = x_i]$$

At first sight, this may seem a rather strange definition, but when giving the matter some thought it is not difficult to see that it boils down to computing what is known as a **weighted average**. First, if asked to compute the average of all x_i, anyone would do the obvious and compute

$$\frac{x_1 + x_2 + \cdots + x_N}{N}$$

In essence, what we're doing is giving an equal weight to the contribution that each x_i has to the average of X. In terms of probabilities, we are interested the *expected occurrence* of each x_i, which is determined by the probability $\mathbb{P}[X = x_i]$. In our example of just computing the average, $\mathbb{P}[X = x_i] = \frac{1}{N}$ for each x_i, and indeed,

$$\frac{x_1 + x_2 + \cdots + x_N}{N} = \sum_{i=1}^{N} x_i \cdot \frac{1}{N} = \sum_{i=1}^{N} x_i \cdot \mathbb{P}[X = x_i]$$

If it turns out that the expected occurrence of x_i is higher than that of, say, x_j, x_i's contribution to the average of X will be higher than that of x_j. In other words, we should weigh x_i more than x_j, which, in turn, is expressed by the probability $\mathbb{P}[X = x_i]$. This explains why we can also speak of a weighted average.

7.2. CLASSICAL RANDOM NETWORKS

The mean vertex degree of an $ER(n,p)$ graph is thus computed as

$$\overline{\delta} \stackrel{\text{def}}{=} \mathbb{E}[\delta] \stackrel{\text{def}}{=} \sum_{k=1}^{n-1} k \cdot \mathbb{P}[\delta = k]$$

We can now prove the following:

Theorem 7.1: *The expected value for the vertex degree of an $ER(n,p)$ graph is equal to $p(n-1)$.*

Proof. To compute the mean vertex degree, we proceed as follows:

$$\begin{aligned}
\sum_{k=1}^{n-1} k \cdot \mathbb{P}[\delta = k] &= \sum_{k=1}^{n-1} \binom{n-1}{k} k\, p^k (1-p)^{n-1-k} \\
&= \sum_{k=1}^{n-1} \frac{(n-1)!}{k!(n-1-k)!} k\, p^k (1-p)^{n-1-k} \\
&= \sum_{k=1}^{n-1} \frac{n-1}{k} \frac{(n-2)!}{(k-1)!(n-k-1)!} k\, p \cdot p^{k-1} (1-p)^{n-1-k} \\
&= p(n-1) \sum_{k=1}^{n-1} \frac{(n-2)!}{(k-1)!(n-k-1)!} p^{k-1} (1-p)^{n-1-k} \\
&= p(n-1) \sum_{l=0}^{n-2} \frac{(n-2)!}{l!(n-l-2)!} p^l (1-p)^{n-l-2} \\
&= p(n-1) \sum_{l=0}^{n-2} \binom{n-2}{l} p^l (1-p)^{n-l-2} \\
&= p(n-1) \sum_{l=0}^{m} \binom{m}{l} p^l (1-p)^{m-l} \\
&= p(n-1) \cdot 1
\end{aligned}$$

□

In other words, our best guess at what the vertex degree of an arbitrarily chosen vertex from an $ER(n,p)$ is, is $p(n-1)$.

We will often use the abbreviation $P[k]$ instead of $\mathbb{P}[\delta = k]$. Let's take a few specific examples of $ER(n,p)$ graphs and analyze some of their properties with the techniques introduced in the previous chapter. In particular, we first consider the case where $n = 100$ and $p = 0.3$. As mentioned, it is important to realize is that there are many different (i.e., nonisomorphic) graphs that qualify as being an $ER(100, 0.3)$ graph. As a consequence, if we are to consider an example, we will have to *construct* a specific $ER(100, 0.3)$ graph. There are various ways to do this, to which we return later, but for now, let G be such a constructed graph. In our case, G is constructed to be connected, simplifying our analysis.

We have just derived an exact expression for the vertex degree distribution of $ER(n,p)$ graphs, which is shown in Figure 7.1(a) as the smooth curve. However, because we are considering one specific random graph, the distribution of vertex degrees for G may differ from this one, as is also shown. The situation changes when considering larger random graphs, as shown in Figure 7.1(b). In general, if we consider increasingly larger graphs with the same expected vertex degree as G, we would see that our specific examples would also better approximate the theoretical degree distribution. To give a hint on why this is so, two observations are important. First, by considering graphs with the same expected vertex degree, we will essentially see that the range of observed vertex degrees is the same for all graphs, independent of their respective size. Second, if the possible vertex degrees are the same, larger graphs will have many more vertices of degree k than smaller graphs. As a consequence, the fluctuation (i.e., standard deviation) that we can expect to see in the number of vertices having degree k will also be smaller. It is beyond the scope of this text to explain these matters in more detail.

Note 7.2 (More information)
These examples illustrate that when analyzing a network, it may sometimes be difficult (if not impossible) to draw the correct conclusion as to what kind of network we're actually dealing with. For example, by just looking at the specific vertex degrees of graph G from Figure 7.1(a), we may not even suspect that we are dealing with an $ER(100, 0.3)$ graph. We could be more confident in the case of the $ER(2000, 0.015)$ graph in Figure 7.1(b), but in both cases we would need to formulate a *statistical test* to draw any real conclusions. In practice, we simply use several metrics to see what kind of graph we're dealing with.

Other metrics for random graphs

Let's consider some other metrics for ER random graphs. First, Fronczak et al. [2004] show that for (large) random graphs $H \in ER(n, p)$, the average path length can be estimated as

$$\bar{d}(H) = \frac{\ln(n) - \gamma}{\ln(pn)} + 0.5$$

where γ is the so-called *Euler constant* (which is approximately equal to 0.5772). We have just seen that the average vertex degree $\bar{\delta}$ for an $ER(n, p)$ graph is equal to $p(n-1)$, which means that for large n, we can also esti-

7.2. CLASSICAL RANDOM NETWORKS

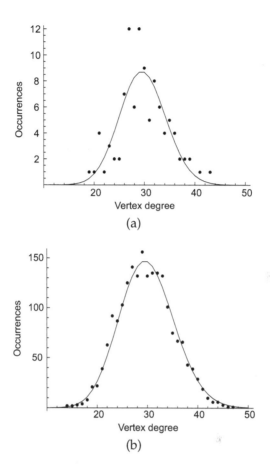

Figure 7.1: Degree distribution of a $ER(100, 0.3)$ random graph, and the values for one specific graph G from that class (a), and similarly for a graph $G^* \in ER(2000, 0.015)$ (b).

mate the average path length as:

$$\bar{d}(H) = \frac{\ln(n) - \gamma}{\ln(\bar{\delta})} + 0.5$$

To give an impression of what this means, Figure 7.2 shows these estimations for different ER graphs. In Figure 7.2(a) we vary the number of vertices, but keep the average vertex degree the same for the differently sized graphs. In this way, we can compare graphs having different *network densities*. In Figure 7.2(b) we show the effect of adding more edges (and thus increasing the network density and average vertex degree) while keeping the number of vertices constant. Clearly, the average path length drops logarithmically.

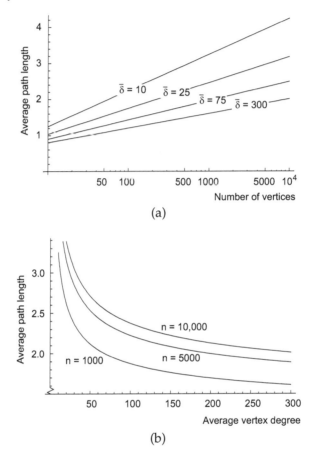

Figure 7.2: The average path length for (a) ER random graphs of different sizes and fixed average vertex degree, and (b) while varying the average vertex degrees for fixed-size graphs. Figure (a) uses a logarithmic x-axis.

What about the clustering coefficient? Recall that the clustering coefficient of a vertex is computed as the fraction of edges found between the neighbors of that vertex, and the maximum number of possible edges between those neighbors. It is not difficult to see that for an $ER(n, p)$ random graph the expected value of the clustering coefficient is equal to p. This can be formally proved as follows.

Theorem 7.2: *The clustering coefficient of any $ER(n, p)$ is equal to p.*

7.2. CLASSICAL RANDOM NETWORKS

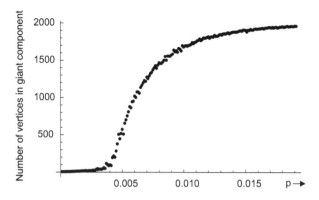

Figure 7.3: The evolution of the size of the giant component in graphs G_p from $ER(2000, p)$ as a function of p.

Proof. Consider an $ER(n, p)$ graph $G_{n,p}$ and an arbitrary vertex $v \in V(G_{n,p})$ with neighbor set $N(v)$. Let $n_v = |N(v)|$. Any two distinct neighbors have probability p to be joined by an edge. Therefore, with $\binom{n_v}{2}$ possible neighbor pairs, we can expect a total of $p \cdot \binom{n_v}{2}$ edges between v's neighbors. The maximum number of edges is equal to $\binom{n_v}{2}$, so that $cc(v) = p \cdot \binom{n_v}{2} / \binom{n_v}{2} = p$. This also means that $CC(G_{n,p}) = p$. □

Indeed, when we consider our example graphs from Figure 7.1, we find that

$$CC(G) = 0.299 \quad \text{and} \quad CC(G^*) = 0.0150$$

When we take a look at the connectivity of a random graph G, we find an interesting relationship between the probability p of connecting two vertices and the size of the components of G. It turns out that when increasing p, not only does the network density of G increase (this should come as no surprise), but also that most vertices are contained in one component while the rest are scattered among a few very small ones. This one component is generally referred to as the **giant component**. The formal mathematics underlying the theory of the size of components in random graphs is beyond the scope of this text. However, it is not difficult to take an experimental approach to observe what is going on by simply constructing graphs with a fixed number of vertices, but changing p. Figure 7.3 illustrates how the size of the giant component evolves in relation to increasing p. In this case, we take a look at the number of vertices in the largest component of a graph $G_p \in ER(2000, p)$ while changing p.

It is interesting to see how quite suddenly the giant component appears: as soon as p comes close to even a small value such as 0.005, we see that the giant component swiftly moves from containing less than 25% of all vertices to a near 100% when $p = 0.015$. In other words, vertices quickly join together in a single component.

Moreover, and in line with this observation, random graphs generally tend to be very well connected. In other words, for a random graph G, the size of the minimal vertex cut $\kappa(G)$ turns out to be fairly large. In fact, in many real-world situations we often have to remove 70-80% of the vertices before the remaining graph partitions into several components. What is remarkable even in that case, is that we will again find most vertices grouped into a single, large component, along with a few, very small components (mostly consisting of just a single vertex). Figure 7.4 shows what happens if we take our example graph G^* from $ER(2000, 0.015)$ and systematically remove vertices. At the same time, we count how many vertices are *not* in the giant component.

What Figure 7.4 shows is that we may need to remove as much as 70% of all vertices before the graph partitions. However, even in that case, we will still find most of the vertices in the same component. Even after having removed 95% of all vertices, half of the remaining vertices will be connected through a path. Note that the fraction of vertices outside the giant component *decreases* when having removed more than 98% of the vertices. (It's not that difficult to figure out why.)

We have just discussed an important feature of random graphs, which emerges from the basic properties of such graphs. As we will see throughout the remainder of the text, many real-world networks combine a large size with randomness, and indeed, most of these networks demonstrate to

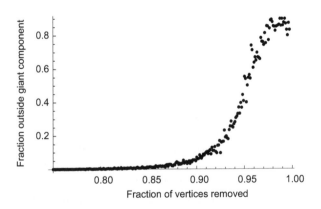

Figure 7.4: The fraction of vertices outside of the giant component when removing vertices from an $ER(2000, 0.015)$ graph.

7.3 Small worlds

In 1967, Stanley Milgram, at that time a professor of social psychology at Harvard, was interested to know what the probability was that two randomly selected people would know each other. This eventually led to the question how far any two persons were separated from each other. Distance was expressed in terms of "A knows B, who knows C, who knows D,...," and so on. In other words, separation was determined by the chain of acquaintances through which one person would eventually reach someone else.

In terms of graphs, Milgram was interested to know the average path length in what is known as a **social network**. In such a network, a vertex represents a person, and an edge between two vertices A and B tells us that A and B are acquaintances. Milgram measured the average path length by asking arbitrary people to send letters to target persons. Let Zach be such a target, and let Alice be a person currently in possession of the letter. If Alice didn't know Zach, she would have to send the letter to one of her acquaintances, say Bob, under the assumption that she would expect Bob to know better than her how to reach Zach. In the original experiment, letters where initially sent from places in the Mid-West of the United States with the targets being located in Massachusetts. Much to his surprise, for those letters that made it to their destination, it took an average of only 5.5 hops, leading to the now famous phrase "six degrees of separation."

What does this have to do with random graphs? What Milgram demonstrated, and what has been shown to hold in many real-world situations, is that the average shortest path length is relatively small. We already saw this to also be true for ER random graphs. However, in many social networks, we also know that people tend to group into relatively small clusters: Alice's acquaintances also know each other. In other words, many social networks (and, in fact, others as well), tend to have a high clustering coefficient.

What we are thus faced with are networks that combine the properties of ER random graphs, yet differ when it comes to the clustering coefficient. Watts and Strogatz [1998] were the first to propose a method to construct such networks, which has since then spawned a wealth of research on constructing and studying similar random graphs, now collectively referred to as **small-world networks**. The procedure proposed by Watts and Strogatz is as follows:

Algorithm 7.1 (Watts-Strogatz): *Consider a set of n vertices $\{v_1, v_2, \ldots, v_n\}$ and* an (even) number k. In order to ensure that the graph will have relatively few edges (i.e., it is sparse), choose n and k such that $n \gg k \gg \ln(n) \gg 1$.

1. Order the n vertices into a ring and connect each vertex to its first $k/2$ left-hand (clockwise) neighbors, and to its $k/2$ right-hand (counterclockwise) neighbors, leading to graph G^1.

2. With probability p, replace each edge $\langle u, v \rangle$ with an edge $\langle u, w \rangle$ where w is a randomly chosen vertex from $V(G)$ other than u, and such that $\langle u, w \rangle$ is not already contained in edge set of (the modified) G.

The resulting graph is called a **Watts-Strogatz random graph** (or simply a **WS graph**). We also refer to a $WS(n, k, p)$ graph.

The notation "$n \gg k$" means that n should be much larger than k. Note that, just as with ER random graphs, $WS(n, k, p)$ actually denotes a collection of graphs. To get a first impression of the effect of changing p, Figure 7.5 shows a small WS graph of only 20 vertices. With $k = 8$, and $\ln(n) \approx 3$, we also see that this example barely meets the conditions for proper WS graphs.

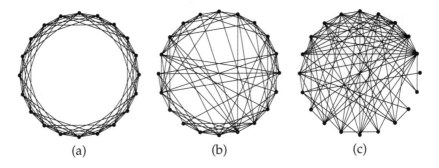

Figure 7.5: Three WS graphs with $n = 20$, $k = 8$, and (a) $p = 0.0$, (b) $p = 0.20$, and (c) $p = 0.90$, respectively.

It is not difficult to see that the maximum distance between any two connected vertices in Figure 7.5(a) is equal to 3. In general, for a Watts-Strogatz graph from $WS(n, k, 0)$, it can be shown that this maximum distance is equal to the smallest integer larger or equal to $(n/2)/(k/2)$ (i.e., $\lceil n/k \rceil$). What Watts and Strogatz establish with their construction is that many vertices will stay close together, but that most vertices will also have a link to a vertex that is relatively far away. In social networks, such a link represents a

---

[1] Recall that we orient a vertex toward the *middle* of the ring in order to give sensible meaning to left- and right-hand.

7.3. SMALL WORLDS

tie between different communities, and as we shall discuss later, these so-called **weak links** play a crucial role in many societies (see also Csermely [2006]). As a consequence, one would expect to see randomness combined with a high degree of clustering.

And indeed, when examining the clustering coefficient for large Watts-Strogatz graphs, it turns out that it stays close to the value we find for the case in which $p = 0$, even for relatively large values of p. If we consider the specific case $p = 0$, we can even prove that the clustering coefficient can be as high as $\frac{3}{4}$. The proof is a bit tedious, but not inherently difficult.

Theorem 7.3: *For any Watts-Strogatz graph G from WS(n, k, 0) the clustering coefficient for G is equal to $CC(G) = \frac{3}{4} \frac{(k-2)}{(k-1)}$.*

Proof (*). In the following, we shall make use of a simple distance metric

$$d_2^n(i,j) \stackrel{\text{def}}{=} \min\{|i-j|, n-|i-j|\}$$

which tells us how far two nodes are when measured along the "outer ring" of a WS graph. For example, in Figure 7.5(a), $d_2^{20}(1,20) = 1$, $d_2^{20}(1,18) = 3$, and so on. We have deliberately used the notation d_2^n to indicate the case $k = 2$ for a $WS(n, k, 0)$ graph. Indeed, a $WS(n, 2, 0)$ graph is nothing but a collection of vertices organized as ring.

Let u be an arbitrary vertex from G and consider the subgraph H induced by its set of k neighbors $N(u)$. $N(u)$ thus consists of u's $k/2$ right-hand (i.e., counter clockwise) neighbors $\{v_1^-, v_2^-, \ldots, v_{k/2}^-\}$ and likewise, its $k/2$ left-hand (i.e., clockwise) neighbors $\{v_1^+, v_2^+, \ldots, v_{k/2}^+\}$ as shown in Figure 7.6.

Consider the degree of vertex v_1^-. The "farthest" right-hand neighbor of v_1^- is $v_{k/2}^-$, where farthest is defined with respect to the distance metric d_2^n. This means that v_1^- has $k/2 - 1$ right-hand neighbors in H. Likewise, v_2^- has $k/2 - 2$ right-hand neighbors, and, in general, v_i^- has $k/2 - i$ right-hand neighbors in H. Clearly, each vertex v_i^- is missing only u as its left-hand neighbor in H, meaning that it has $k/2 - 1$ left-hand neighbors. We can therefore conclude that the vertex degree of v_i^- is equal to

$$\delta(v_i^-) = \left(\frac{k}{2} - i\right) + \left(\frac{k}{2} - 1\right) = k - i - 1$$

A completely analogous reasoning holds for all vertices v_i^+, so that $\delta(v_i^+) =$

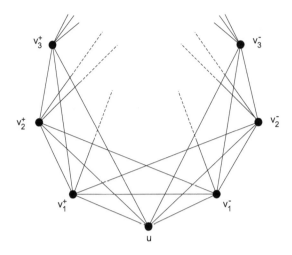

Figure 7.6: The labeling of vertices left and right of u.

$k - i - 1$. As a consequence, the total number of edges in H is equal to

$$\begin{aligned}|E(H)| &= \tfrac{1}{2} \sum_{v \in V(H)} \delta(v) \\ &= \tfrac{1}{2} \sum_{i=1}^{k/2} \left(\delta(v_i^-) + \delta(v_i^+) \right) \\ &= \tfrac{1}{2} \sum_{i=1}^{k/2} \left((k-i-1) + (k-i-1) \right) = \sum_{i=1}^{k/2} (k-i-1) \end{aligned}$$

Knowing that $\sum_{i=1}^{m} i = \tfrac{1}{2} m(m+1)$, we obtain

$$\begin{aligned} |E(H)| &= \sum_{i=1}^{k/2} (k-i-1) = k/2(k-1) - \sum_{i=1}^{k/2} i \\ &= k/2(k-1) - \tfrac{1}{2}(k/2)(k/2+1) \\ &= \tfrac{3}{8} k(k-2) \end{aligned}$$

Because $|V(H)| = k$, we compute the clustering coefficient $cc(u)$ for vertex u as

$$cc(u) = \frac{|E(H)|}{\binom{|V(H)|}{2}} = \frac{\tfrac{3}{8}k(k-2)}{\tfrac{1}{2}k(k-1)} = \frac{3}{4} \frac{(k-2)}{(k-1)}$$

Because all vertices are the same in G, $CC(G) = cc(u)$, completing the proof. □

7.3. SMALL WORLDS

As we mentioned, the idea behind Watts-Strogatz graphs is to combine properties of classical random graphs with high clustering coefficients. What the previous theorem tells us is that the clustering coefficient of a $WS(n,k,0)$ graph is independent of its size and that for large values of k it is close to $\frac{3}{4}$.

As we already saw, a characteristic property of ER random graphs is the relatively short average path length. For a $WS(n,k,0)$ graph, however, it is not difficult to see that the average shortest-path length between two vertices may be relatively long. For example, for a $WS(n,k,0)$ graph we have the following theorem.

Theorem 7.4: *For a Watts-Strogatz graph G from $WS(n,k,0)$ the average shortest-path length $\bar{d}(u)$ from a given vertex u to any other vertex in G is approximated by*

$$\bar{d}(u) \approx \frac{(n-1)(n+k-1)}{2kn}$$

Proof. Consider a given vertex u. Using the same notation as before, let $L(u,1)$ be the $k/2$ left-hand neighbors $\{v_1^+, v_2^+, \ldots, v_{k/2}^+\}$ of vertex u. Likewise, let $L(u,2) = \{v_{k/2+1}^+, \ldots, v_k^+\}$ be the set of $k/2$ left-hand *next* neighbors, i.e., neighbors of u at distance 2. In general, we have that $L(u,m)$ is the set of vertices $\{v_{(m-1)k/2+1}^+, \ldots, v_{mk/2}^+\}$ left of u connected to u by a (shortest) path of length m. Another way to see this is noting that each vertex in $L(u,m)$ is connected to a vertex from $L(u,m-1)$. $L(u,m)$ thus consists of all left-hand neighbors of u at (shortest) distance m. Similarly, we can define the sets $R(u,m)$ of right-hand neighbors at distance m. Note also that the index p of the farthest vertex v_p^+ contained in any $L(u,m)$ will be less than approximately $(n-1)/2$, which is roughly at the other end of the ring along which the vertices have been organized. Because all sets $L(u,m)$ have equal size, this also means that $m \leq \frac{(n-1)/2}{k/2}$. This gives us:

$$\bar{d}(u) \approx \frac{1}{n} \sum_{i=1}^{(n-1)/k} \left(i \cdot |L(u,i)| + i \cdot |R(u,i)| \right) \approx \frac{1}{n} \sum_{i=1}^{(n-1)/k} \left(i \cdot \frac{k}{2} + i \cdot \frac{k}{2} \right)$$

where $i \cdot |L(u,i)|$ is nothing else but the cumulative length of the shortest paths to u's left-hand vertices, and likewise, $i \cdot |R(u,i)|$ the cumulative length of such paths to u's right-hand vertices. This then leads to

$$\bar{d}(u) \approx \frac{k}{n} \sum_{i=1}^{(n-1)/k} i = \frac{k}{2n}\left(\frac{n-1}{k}\right)\left(\frac{n-1}{k}+1\right) = \frac{(n-1)(n+k-1)}{2kn}$$

and completes the proof. □

What does this mean? It tells us that $WS(n,k,0)$ graphs may show a high clustering coefficient, yet miss the property of small worlds, that is, having small average shortest-path lengths. However, it turns out that by only slightly increasing p, the average path length of a Watts-Strogatz graph drops rapidly. On the other hand, the clustering coefficient stays relatively high except when p becomes large as well. These two effects are illustrated in Figure 7.7. In this case, we have examined a range of $WS(1000, 30, p)$ graphs, varying p from very small to relatively large. We compute the clustering coefficients $CC(G)$, but normalize each one of by division through the clustering coefficient for the case that $p = 0$. Likewise, we compute the average path lengths $\bar{d}(G)$, again normalized by division through the value in case $p = 0$. What Figure 7.7 shows is that when increasing p, the average path length drops rapidly, but the clustering coefficient stays relatively high.

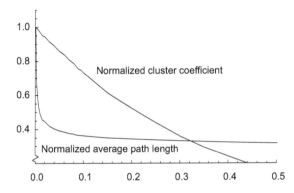

Figure 7.7: The relation between clustering coefficient and average path length for increasing value of p in a $WS(1000, 30, p)$ graph.

7.4 Scale-free networks

The Watts-Strogatz model of networks is generally considered to represent small-world phenomenon. However, WS random graphs often do not capture (other) properties of real-world networks, such as communication networks or biological networks. It was the work by Albert-László Barabási and his student Réka Albert which caused an avalanche of research on so-called **scale-free networks**. Roughly speaking, they showed that real-world networks such as the World Wide Web, actor collaborations, and many more, exhibit a structure in which there are a few high-degree nodes, but that the number of nodes with a high degree decreases exponentially [Barabási and

7.4. SCALE-FREE NETWORKS

Albert, 1999]. In this section, we will take a closer look at this phenomenon (and will also be more precise in our formulation).

Fundamentals

By now, it has become common practice to call a network scale free if the distribution of vertex degrees follows a **power law**. Roughly speaking, this means that the probability that an arbitrary node has degree k is proportional to $(1/k)^\alpha$ for some number $\alpha > 1$ called the **scaling exponent**. In mathematical terms, $P[k] \propto k^{-\alpha}$.

For most real-world scale-free networks, it turns out that $2 < \alpha < 3$. As an example, consider an artificially constructed scale-free network G with 2000 nodes. Figure 7.8 shows the degree distribution of G. For clarity, we show the distribution in two different ways. Figure 7.8(a) gives us the usual way of displaying relationships, namely using linear scales for the x and y axes. In Figure 7.8(b) we show the same results, but now using a logarithmic scale for both axes. In this case, we essentially see a straight line. To be more specific, using a curve-fitting method as briefly described in Note 6.1 on page 136, one can show that

$$P[k] \approx 324 \cdot x^{-0.62} + 2.3$$

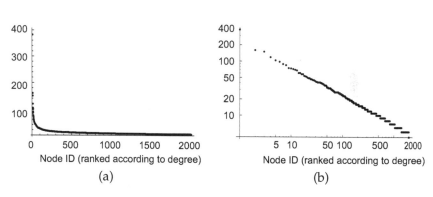

Figure 7.8: The distribution of vertex degrees of a scale-free network with 2000 nodes, shown as (a) a linear plot and as (b) a log-log plot.

The network from Figure 7.8 contains 2000 nodes with a median vertex degree of 7. In other words, half of the vertices have a degree of 7 or less. Interestingly enough, the highest-degree node is connected to no less that 382 other nodes, whereas the second-highest vertex degree is 160. This occurrence of a few **hubs** is typical for scale-free networks. To complete the picture, in our example network only 10% of the nodes have a degree larger than 50.

Note 7.3 (More information)
To understand why such networks are called scale-free, we note that formally a function $f(x)$ is called scale-free when it satisfies the following property:

$$f(bx) = C(b) \cdot f(x)$$

where $C(b)$ is some constant dependent only on b. The basic idea is that the overall form of the function f does not change when considering values for x that are a factor b larger. As it turns out, power-law distributions obey this property, i.e., if $f(x) = x^{-\alpha}$, we find that

$$f(bx) = (bx)^{-\alpha} = b^{-\alpha} \cdot x^{-\alpha} = b^{-\alpha} f(x)$$

This can be nicely illustrated by our example scale-free graph G from Figure 7.8. Figure 7.9 shows the degree distribution for nodes ranked between position 10 and 100, and between 100 and 1000, respectively. What is immediately clear is that the *form* of the degree distribution is almost the same, i.e., independent of the range of rankings we consider. This aspect is characteristic for scale-free distributions.

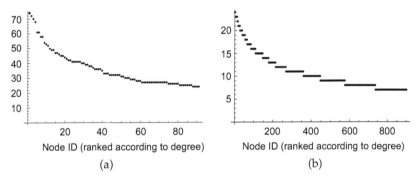

Figure 7.9: The degree distribution of nodes ranked between (a) 10 and 100, and (b) between 100 and 1000.

ER random networks have been defined as graphs where there is a probability that two vertices are adjacent. Watts-Strogatz networks are constructed by *rewiring* edges, that is, changing a well-structured graph by probabilistically repositioning its current edges between different vertices regardless the degree of the original end points. As explained by Dorogovtsev et al. [2003] and Vega-Redondo [2007], scale-free graphs are fundamentally different because it appears that we can construct them only through a

7.4. SCALE-FREE NETWORKS

growth process combined with what is referred to as **preferential attachment**. In other words, to understand the structure of real-world networks (which are generally scale free), we need to concentrate on how they have come to existence by observing how new nodes attach themselves to existing nodes.

Barabási and Albert [1999] were the first to devise a procedure for the construction of scale-free networks. Their procedure combines the growing of a network with attaching new nodes to existing ones with certain preferences. The algorithm is as follows:

Algorithm 7.2 (Barabási-Albert): *Consider a (relatively small) ER random graph G_0 with n_0 vertices V_0. At each step $s > 0$:*

1. *Add a new vertex v_s to V_{s-1} (i.e., $V_s \leftarrow V_{s-1} \cup \{v_s\}$).*

2. *Add $m \leq n_0$ edges to the graph, each edge being incident with v_s and a vertex u from V_{s-1} chosen with probability*

$$\mathbb{P}[select\ u] = \frac{\delta(u)}{\sum_{w \in V_{s-1}} \delta(w)}$$

   that is, choosing a vertex u is proportional to the current vertex degree of u. Vertex u must not have been previously chosen during this step.

3. *Stop when n vertices have been added, otherwise repeat the previous two steps.*

*The resulting graph is called a **Barabási-Albert random graph**, or simply a **BA graph**. We also refer to a $BA(n, n_0, m)$ graph.*

Obviously, after t steps we will have a graph with $t + n_0$ vertices and $t \cdot m + |E(G_0)|$ edges. (Note that G_0 may have no edges to begin with.) Barabási and Albert [1999] show that for this model, the probability $P[k]$ that an arbitrary vertex v has degree k is proportional to k^{-3}.

Note 7.4 (More information)
To get a better grasp on why the degree distribution of a BA graph is proportional to k^{-3}, we adopt the notations and approach as found in [Vega-Redondo, 2007] (and which were originally introduced by Dorogovtsev et al. [2000]). Formally, we have the following:

Theorem 7.5: *For any $BA(n, n_0, m)$ graph G, the probability that vertex $v \in V(G)$ has degree $k \geq m$ is given by:*

$$P[k] = \frac{2m(m+1)}{k(k+1)(k+2)} \propto \frac{1}{k^3}$$

Proof (*). Let $q_t(s,k)$ denote the probability that at step t vertex v_s has degree k (with $s < t$). In order for the degree of v_s to increase by 1, it is necessary that v_t attaches to v_s. There are m opportunities to let this happen, each with probability $\mathbb{P}[\text{select } u]$ as given above. If we assume that $|E(G_0)| = 0$, we know that there are a total of $m(t-1)$ edges just before step t so that $\sum_{w \in V_{t-1}} \delta(v) = 2|E| = 2m(t-1)$. In other words, the probability that v_s will be attached to v_t is

$$\mathbb{P}[\text{attach to } v_s] = m \cdot \frac{\delta(v_s)}{\sum_{w \in V_{s-1}} \delta(w)} = \frac{m(k-1)}{2m(t-1)} = \frac{k-1}{2(t-1)}$$

The probability that the degree of v_s was k and stays so, is equal to $1 - \frac{k}{2(t-1)}$. Combining these two results, it should then be clear that

$$q_t(s,k) = \left(\frac{k-1}{2(t-1)}\right) \cdot q_{t-1}(s, k-1) + \left(1 - \frac{k}{2(t-1)}\right) \cdot q_{t-1}(s,k) \quad (7.1)$$

The first term represents the situation where the new node attaches to s, whereas the second term covers the case where s's degree stays the same (namely k).

We are interested in finding the distribution $P_t[k]$ of the vertex degrees after t steps. In other words, we want to know the probability that any vertex $v_1, \ldots, v_t$ has degree k. The probability that vertex s has degree k at step t is largely independent of that of vertex $s' > s$, certainly when s is large. This means that we can simply compute $P_t[k]$ as

$$P_t[k] = \overline{q_t(\cdot, k)} = \frac{1}{t} \sum_{s=1}^{t} q_t(s,k)$$

Using expression (7.1) we need to distinguish two cases. First, If $k > m$ we know that the added vertex v_t does not belong to the set of vertices with degree k, so that we have

$$\sum_{s=1}^{t} q_t(s,k) = \frac{k-1}{2(t-1)} \sum_{s=1}^{t-1} q_{t-1}(s, k-1) + \left(1 - \frac{k}{2(t-1)}\right) \sum_{s=1}^{t-1} q_{t-1}(s,k)$$

However, if $k = m$, it must be the case that v_t is in this set as well. In other words,

$$\sum_{s=1}^{t} q_t(s,m) = \frac{m-1}{2(t-1)} \sum_{s=1}^{t-1} q_{t-1}(s, m-1) + \left(1 - \frac{m}{2(t-1)}\right) \sum_{s=1}^{t-1} q_{t-1}(s,m) + 1$$

To keep matters simple, we will first concentrate on the situation that $k > m$. We are seeking to express $q_t(s,k)$ in terms of the probability $P_t[k]$. By straight-

7.4. SCALE-FREE NETWORKS

forward algebraic manipulation, we obtain the following:

$$\sum_{s=1}^{t} q_t(s,k) = \frac{1}{2}\frac{t-1}{t-1}(k-1)\left(\frac{1}{t-1}\sum_{s=1}^{t-1} q_{t-1}(s,k-1)\right)$$
$$- \frac{1}{2}\frac{t-1}{t-1}k\left(\frac{1}{t-1}\sum_{s=1}^{t-1} q_{t-1}(s,k)\right)$$
$$+ (t-1)\left(\frac{1}{t-1}\sum_{s=1}^{t-1} q_{t-1}(s,k)\right)$$
$$= \frac{1}{2}\Big((k-1)P_{t-1}[k-1] - kP_{t-1}[k]\Big) + (t-1)P_{t-1}[k]$$

Knowing that

$$\sum_{s=1}^{t} q_t(s,k) = t\left(\frac{1}{t}\sum_{s=1}^{t} q_t(s,k)\right) = tP_t[k]$$

and $\lim_{t\to\infty} P_t[k] = P[k]$, we find

$$(k+2)P[k] - (k-1)P[k-1] = 0$$

When we consider the special case $k = m$, we apply exactly the same algebraic manipulations to find that

$$(m+2)P[m] - (m-1)P[m-1] = 2$$

Of course, $P[m-1] = 0$, as there can be no vertex with a degree lower than m, which means that $P[m] = 2/(m+2)$. We now have:

$$P[k] = \frac{k-1}{k+2}P[k-1] = \frac{k-1}{k+2}\frac{k-2}{k+1}P[k-2] = \ldots = \frac{m(m+1)(m+2)}{k(k+1)(k+2)}P[m]$$

Substituting $P[m]$ in this equation gives us

$$P[k] = \frac{2m(m+1)}{k(k+1)(k+2)}$$

which completes the proof. □

It should be mentioned that BA graphs are not the only ones for constructing scale-free networks. One particular interesting extension to the Barabási-Albert model is the following:

Algorithm 7.3 (Generalized Barabási-Albert): *Consider a small graph G_0 with n_0 vertices V_0 and no edges. At each step $s > 0$:*

1. *Add a new vertex v_s to V_{s-1}.*

2. Add $m \leq n_0$ edges, each edge being incident to v_s and a vertex u from V_{s-1} chosen with probability proportional to its current degree $\delta(u)$ (and not previously chosen in this step).

3. For some constant $c \geq 0$ add another cm edges between vertices from V_{s-1}, where the probability of adding an edge between vertices u and w is proportional to the product $\delta(u) \cdot \delta(w)$, and under the condition that $\langle u, w \rangle$ does not yet exist.

4. Stop when n vertices have been added.

As shown by Dorogovtsev et al. [2003], the resulting graph corresponds to a scale-free network for which the vertex degree is proportional to

$$P[k] \propto k^{-(2+\frac{1}{1+2c})}$$

In other words, for $c = 0$ we have a BA graph, but for increasing values of c, the exponent converges to 2.

Properties of scale-free networks

As it may have become clear by now, formal analysis of random networks is generally far from trivial. This is certainly also true for the scale-free networks discussed previously. The consequence of this observation is that in order to attain insight in the properties of scale-free networks, we need to simply apply the network analysis tools from Chapter 6 and see what we can learn from experiments.

Let us first consider the clustering coefficient. As we have seen for ER random graphs, the clustering coefficient can be expressed independently of the size of the graph. For Watts-Strogatz graphs, we have shown that the clustering coefficient is large and stays almost the same even for relatively large rewiring probabilities. More importantly is that for Watts-Strogatz graphs the clustering coefficient is independent of the number of vertices.

The situation for scale-free networks is more complicated. In fact, finding an analytical expression that estimates the clustering coefficient for general scale-free networks has not yet been found. Fronczak et al. [2003] considered the situation for BA random graphs, and, in particular, looked at the clustering coefficient $cc(v_s)$ of vertex v_s after t steps had taken place in the construction of a $BA(t, n_0, m)$ graph (of course, $s \leq t$). They find:

$$cc(v_s) = \frac{m-1}{8(\sqrt{t} + \sqrt{s}/m)^2} \left(\ln^2(t) + \frac{4m}{(m-1)^2} \ln^2(s) \right)$$

When evaluating this somewhat ghastly expression for fixed values of m and t, yet varying s, we obtain low clustering coefficients, as shown in Figure 7.10.

7.4. SCALE-FREE NETWORKS

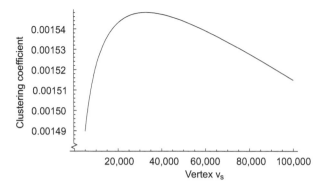

Figure 7.10: The clustering coefficient for vertices v_s in a $BA(100\,000, n_0, 8)$ graph.

To see how these values compare to those of an ER random graph, we consider an ER random graph with the same number of vertices and the same average vertex degree. We first compute the average vertex degree of a $BA(n, n_0, m)$ graph for very large n. As proven in Note 7.4, the degree distribution for a BA graph is given by

$$P[k] = \frac{2m(m+1)}{k(k+1)(k+2)}$$

Taking exactly the same approach for computing the average vertex degree for an ER random graph, the average vertex degree for a BA random graph can be computed as:

$$\bar{\delta}(G) = \mathbb{E}[k] = \sum_{k=m}^{\infty} k \cdot P[k] = 2m(m+1) \sum_{k=m}^{\infty} \frac{k}{k(k+1)(k+2)} = 2m$$

(We leave it as an exercise to the reader to show that this computation of $\mathbb{E}[k]$ is indeed correct.) For a vertex of an ER random graph, we know that $cc(v) = p$ and that $\bar{\delta}(v) = p(n-1)$. In other words, to get the same average vertex degree for an ER random graph as that of a BA graph, we need to take p equal to $2m/(n-1)$. For our example from Figure 7.10, we then find that $cc(v) = 16/9999 \approx 0.00016$. This means that roughly speaking, the clustering coefficient in BA graphs is an order higher than that of ER graphs, yet it remains relatively small. Considering that many real-world networks combine scale-freeness *and* high clustering, it is clear that BA graphs do not form an adequate model of real life. We return to this issue shortly.

What about average path lengths? Fronczak et al. [2004] derive the following estimation of the average path length for a $BA(n, n_0, m)$ random graph:

$$\bar{d}(BA) = \frac{\ln(n) - \ln(m/2) - 1 - \gamma}{\ln(\ln(n)) + \ln(m/2)} + 1.5$$

where γ is the *Euler constant*, which we also came across when estimating the average path length for ER random graphs. To get a better idea of what this estimation means, we can make a comparison with ER random graphs. To this end, consider ER and BA random graphs having the same average vertex degree, and compare their respective average path lengths. The result for $\bar{\delta} = 10$ is shown in Figure 7.11 (and again using a linear and a logarithmic scale for the x axis).

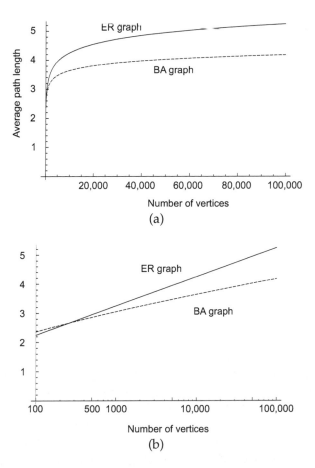

Figure 7.11: Comparing the average path length of ER and BA random graphs with the same average vertex degree on a (a) linear plot and (b) a log-linear plot.

7.4. SCALE-FREE NETWORKS

What is illustrated in this figure is that BA graphs tend to systematically have a relatively much lower average path length than ER random graphs. Considering that the average path length for random graphs is already very low, this is a somewhat remarkable result. On the other hand, unlike ER random graphs, we are now dealing with graphs containing hubs: vertices with high degrees, essentially acting as intermediates between other, less well-connected vertices. For example, one may expect that the eccentricity of a hub is relatively low: a hub is simply close to every vertex. But this also means that most vertices can easily reach other by means of a path containing a hub.

We are thus dealing with what are also called **super small worlds**. And although being able to reach another vertex in only a few steps is a nice property of a large graph, the hubs do form a potential bottleneck. In communication networks, they would generally need to process a lot of transient traffic. Worse is that they may also be vulnerable to attacks. Intuitively, systematically disabling hubs should quickly partition a network into several disjoint components, a highly undesirable situation.

To illustrate these matters, Figure 7.12 shows what happens when we systematically remove vertices from a scale-free graph in comparison to removing the best-connected vertices from an ER random graph. We also show the effect of removing randomly selected vertices from a scale-free graph (which is very similar to randomly removing vertices from an ER graph). A scale-free network is thus seen to be sensitive to a targeted attack, but just as robust as an ER random graph in the case of a random attack.

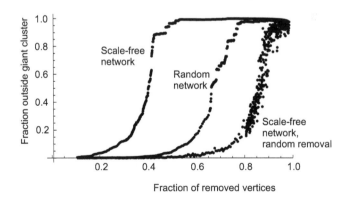

Figure 7.12: The fraction of vertices outside the giant component when removing hubs from a scale-free graph, and those from an ER random graph.

Related networks

As we mentioned, the Barabási-Albert approach for constructing a scale-free graph has one important shortcoming when comparing it to real-world networks: its relatively low clustering coefficient. A better understanding of real-world phenomena should normally be reflected by better models and in this sense, a BA random graph is difficult to validate against many real-world data. Therefore, researchers have been seeking solutions for constructing scale-free graphs that have a high clustering coefficient.

As argued by Dorogovtsev et al. [2003], constructing such graphs is actually quite simple. The trick is to make sure that there are many triangles. This can be achieved, for example, by adding an edge to a triple at each step of the growing process. (Recall that a triple was a subgraph with 3 vertices and 2 edges.) Holme and Kim [2002] provide a scheme that combines scale-freeness and at the same time allows to tune to what extent clustering is to be provided. Their algorithm proceeds as follows:

Algorithm 7.4 (Barabási-Albert with tunable clustering): *Consider a small graph G_0 with n_0 vertices V_0 and no edges. At each step $s > 0$:*

1. *Add a new vertex v_s to V_{s-1}.*

2. *Select a vertex u from V_{s-1} that is not adjacent to v_s and with a probability proportional to its degree $\delta(u)$. Add edge $\langle v_s, u \rangle$. Add the remaining $m - 1$ edges as follows:*

    a) *If $m - 1$ edges have been added, continue with Step 3. Otherwise, proceed with the next step.*

    b) *With probability q: select a vertex w that is adjacent to u, but not to v_s. If no such vertex exists, continue with Step 2c. Otherwise, add edge $\langle v_s, w \rangle$ and continue with Step 2a.*

    c) *Select a vertex u' from V_{s-1} that is not adjacent to v_s and with a probability proportional to its degree $\delta(u')$. Add edge $\langle v_s, u' \rangle$ and set $u \leftarrow u'$. Continue with Step 2a.*

3. *Stop when n vertices have been added, otherwise repeat from Step 1.*

What happens in this approach, is that with probability q we explicitly construct a triangle between the newly added vertex v_s, the vertex u to with it attaches, and one of u's neighbors w. Intuitively, it should be clear that we are more or less controlling the clustering coefficient of vertex v_s. For example, if we choose $q = 1$, and under the assumption that u has $k \leq m$ neighbors $w_1, w_2, \ldots, w_k$, v_s will connect to u as shown in Figure 7.13. From Chapter 6, where we examined the situation that none of the vertices w_i

7.4. SCALE-FREE NETWORKS

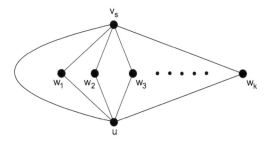

Figure 7.13: The subgraph in which a newly added vertex is contained when attaching to vertex u.

were adjacent to each other, we know that the clustering coefficient for u and v_s is high (and which will grow if edges $\langle w_i, w_j \rangle$ exist).

Holme and Kim [2002] show that their approach yields graphs in which the distribution of the vertex degree follows a power law with scaling exponent $\alpha = 3$. Although they do not derive an analytical expression for the clustering coefficient, experiments show that by varying q, clustering can easily be varied between the one observed for pure BA random graphs, and high values such as 0.5.

Chapter 8

Modern computer networks

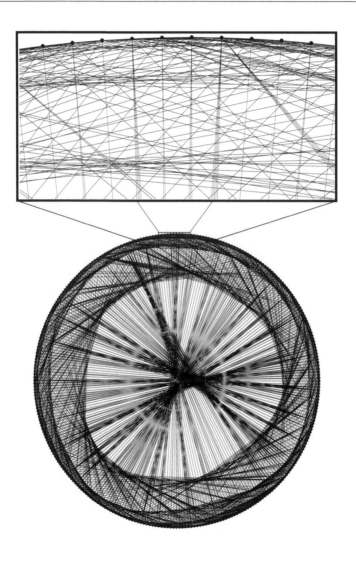

Modern life is difficult to imagine without the Internet. What started in the late 1960s as a simple network of a handful of computers has now grown into an immensely complex communication infrastructure with hundreds of millions of computers and which continues to grow. The Internet as a computer network is often taken to be the same as the World Wide Web (or just simply Web), yet they are fundamentally different. In this chapter we will start with first taking a look at computer networks, in particular the Internet. Second, we'll dive a bit into what are known as overlay networks. These networks are characterized by the fact that a (often very large) group of computers maintain their own communication network and as such form a special type of subnetwork using the Internet as their foundation. Thirdly, we'll pay attention to the World Wide Web and explain where and how it differs from the Internet.

8.1 The Internet

The Internet as a communication network consists of a huge collection of computers connected to each other. The organization of the Internet essentially follows a hierarchical structure consisting of home networks, computer networks in organizations, networks that are owned by Internet Service Providers, and backbone networks, among other types of computer networks. They are all connected together, often using the same infrastructure as used for telephony. Connections may occur through guided media (i.e., wires), but we are increasingly seeing wireless connections for communication as well. In addition, the communication devices vary tremendously: ultra-small networked sensors, smartphones, laptop computers and workstations, servers, routers, and supercomputers. One may wonder how it is even possible to say anything sensible about the structure of the Internet? To answer this question, let's first consider some of the basics and then move onto the phenomenon of interconnected networks.

Computer networks

Small-area networks

There are different ways of characterizing networks, but one that is convenient for our discussion here is simply looking at the physical diameter of a computer network. Typically, networks that span areas up to at most, say, a few hundred meters are characterized by a relatively high density of networked computers, also referred to as **hosts**. Hosts send **packets** to each other through the network that connects them. These networks differ from ones that span large areas, in the sense that **routing** plays a less prominent role. Routing a packet from a source host A to its destination host B means that the packet is required to follow a communication path from A to B. Typically, such paths are set up using one of the shortest path algorithms we discussed in Chapter 5. Without going into further details, setting up or finding a route in a small-area network is relatively easy. Moreover, these small-area networks are generally owned and managed by a single administrative organization.

To get an impression of what we're dealing with, Figure 8.1 shows the typical organization of a small-area network. Such a network consists of several **local-area networks**, or **LAN**s, each typically being a collection of 10-100 computers connected by means of what is known as a **switch**. The switch ensures that a packet addressed to one of its connected computers is forwarded to that computer.

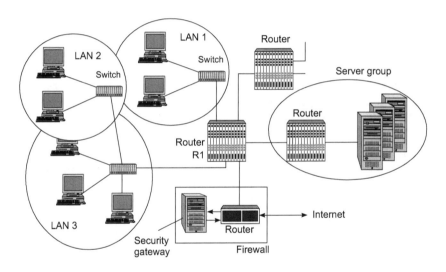

Figure 8.1: A typical example of a small-area network, consisting of a collection of connected local-area networks.

Addresses

LANs can be connected to each other by directly connecting their respective switches, effectively leading to a larger LAN. In addition, it is common practice to use connect LANs through internal **routers**, which we will explain shortly. What is important for our discussion is that each networked host has an **address**. Having an address allows us to send data packets from one

8.1. THE INTERNET

host to another. If we concentrate on the most common case for modern networks, there are two types of addresses we need to distinguish. First, each host has a world-wide unique identifier in the form of a 48-bit number. This so-called **MAC address** comes with the host when it is manufactured (or, more precisely, is associated to a host's network hardware). When a host is connected to a port of a switch (see Figure 8.2), the switch can automatically discover the host's MAC address to subsequently uniquely associate the specific port with that address. As a consequence, when a host with MAC address $MA1$ (connected to port $P1$) requests a packet to be forwarded to host $MA2$ (connected to port $P2$), the switch uses the port identifiers to forward the packet from port $P1$ to $P2$, and thus implicitly from address $MA1$ to address $MA2$.

Figure 8.2: A 16-port switch as used in local-area networks.

More important, however, is the fact that a host can be assigned an **IP address**, where IP stands for **Internet Protocol**. Unlike a MAC address which is *persistent*, meaning that it cannot be changed, an IP address needs to be explicitly assigned when a host is connected to a network. Address assignment can be done manually or automatically, and can be done statically or dynamically. For example, in some cases a separate address assignment service is used to hand out IP addresses with an associated lease time. When a lease expires, the host will need to get a new IP address[1].

A host with IP address $IA1$ normally uses that address to send a packet to a destination, say a host with IP address $IA2$. In contrast to MAC addresses, an IP address can be used to truly route packets through a communication network. In this case, **routers** are represented as the nodes of such a network, and physical links between routers as its edges. In essence, whenever a host wants to send a packet, it needs to make sure that the packet gets to a router, who will then take care of the rest. To this end, it simply sends the packet using the MAC address of a locally accessible router as its destination. From there on, it's the router's job to forward the packet toward its destination.

---

[1]The mechanism just described is generally implemented by means of a so-called **DHCP server**, where DHCP stands for *Dynamic Host Configuration Protocol*.

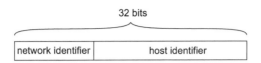

Figure 8.3: The structure of an IP address, consisting of a network identifier and a host identifier.

To avoid that routers need to discover routes to every individual host, a simple aggregation takes place by splitting an IP address into two parts: a **network identifier** and a **host identifier** as shown in Figure 8.3. In the following we will not distinguish among the different types of IP addresses and consider only the ones that are made up of 32-bit numbers. We assume that 16 bits have been reserved for the network identifier and 16 for the host identifier. This means that there can be at most $2^{16} = 32,768$ different networks, each having at most 2^{16} hosts. Whenever a company wants to create a network, it needs to be assigned one or several network identifiers. These identifiers are assigned by a global organization, and will therefore need to be requested. Stepping over many practical matters, in our example network from Figure 8.1, we would need at least three network identifiers: one for the server group, one for LAN #1, and one for the connected LANs #2 and #3. When taking routing decisions, a router considers only the network address and completely ignores the host identifier. So, for example, when router $R1$ from Figure 8.1 receives a packet addressed to a host on LAN #2, it only takes a look at the network identifier in that address and subsequently forwards the packet to the switch of LAN #3, who will then take over the responsibility of getting that packet to its destination.

It turned out that the total number of available network identifiers in the Internet was not enough to support its growth. Therefore, alternative schemes and technical solutions are being used to ensure that each host can be assigned an IP address. Nevertheless, the basic approach just described, namely that each host is addressed by means of a pair of $<network,host>$ identifiers has been left unaltered. This observation is important as routers take decisions on where to forward packets to using only network identifiers.

Other small-area networks

Besides these small-area networks, there are two other types of networks worth mentioning. The first one is formed by **home networks**, which typically consist of one to several end-user computers, along with networked devices such as set-top boxes for digital TV, Internet-enabled telephones, and multimedia centers. These type of networks are growing fast in terms of what they offer to end users. Typically, we are seeing that many domestic appliances are becoming network aware, if alone to smoothly regulate energy consumption. In addition, many home networks facilitate installation of sensors for monitoring purposes (think of burglar systems, networked smoke and fire detectors, surveillance cameras, and so forth). A home network generally has only a single IP address associated with it, which is subsequently shared between all the devices. It is beyond the scope of this text to explain how this sharing is realized. What is important is that a home network from the outside is often indistinguishable from a single networked computer: both have a globally unique IP address.

Secondly, there are also **(wireless) access networks**, whose sole purpose is to allow devices to connect to the Internet. Typically, access networks support wireless connection setups to mobile devices. When making use of such a network, a device is usually provided with a dynamically assigned IP address whose network identifier is inherited from the access network. By keeping track of which device was assigned which IP address, packets are routed to the access network from where a router or switch can forward the packet to its destination.

Large-area networks

Small-area networks form what is known as the **edge of the Internet**: networks beyond which packets are no longer forwarded. In practice, we see these small-area networks be connected to larger networks owned by organizations who make it their business to provide many end users and organizations access to the Internet, or which offer the services to transmit packets across the Internet. These **Internet Service Providers**, or simply **ISPs**, generally span much larger geographical areas than small-area networks. In contrast to the small-area networks discussed previously, routing plays an important role.

The smallest large-area networks consist of the access networks we just discussed (and in this sense, there is usually not a clear-cut distinction between small and large-area networks). Examples include modern wireless access networks that span a whole neighborhood or even a city. In addition, there are many local ISPs that not only provide Internet access, but also basic services such as e-mail.

These so-called **tier-3 networks** have what is known as a **peering relationship** with **tier-2 networks**. A peering relationship between networks $N1$ and $N2$ may occur when $N1$ has a router that is connected through a direct link with a router of $N2$. Such routers are also known as **border gateways**, as they allow for traffic to flow into and from the network, that is, they operate at the border of a network. Tier-2 networks are often connected to other Tier-2 networks, allowing packets to cross larger areas. As said, routing plays a prominent role in these cases. Regional ISPs, such as those covering a (small) country are typical examples of tier-2 networks.

Finally, we distinguish **tier-1 networks**, which provide the backbone of the Internet. End users usually never connect directly to tier-1 networks. Instead, these backbones provide services and routing capabilities only to tier-2 networks. Note that there may be several tier-1 networks operating in the same area. This allows regional ISPs to choose from which network they will make use. In fact, ISPs may change their peering relationships without end users even noticing.

Measuring the topology of the Internet

All of the networks we discussed so far are usually each managed by a separate administrative unit. This is certainly the case for large-area networks. For small-area networks, we often see that the networks are still managed separately (as is typically the case for corporate local-area networks), or management is partly delegated to end users (as with home networks). Roughly speaking, a collection of networks that fall under the regime of the same administration and that follow the same policy regarding how to route packets, is known as an **autonomous system** or simply **AS**. By connecting autonomous systems, we essentially obtain the structure of the Internet. In other words, the Internet can be represented as a graph where a vertex represents an autonomous system, and an edge the fact that two autonomous systems have a peering relationship. As of this writing, there are more than 25,000 autonomous systems.

The AS topology

Discovering what is known as the AS topology of the Internet is on the surface relatively easy provided certain details are not taken into account (and which we will indeed skip for now). Each autonomous system is assigned a unique number called its **AS number**. Note that this assignment is done through a central authority, as is the case for assigning network addresses. Each AS announces which networks fall under its regime by essentially advertising $\langle AS\ number, network\ identifier \rangle$ pairs. Such announcements are made by the AS's border gateways discussed previously, and are picked up by the respective neighboring border gateway of an adjacent AS. As an example, assume that AS 1 manages a network with identifier nid. A border gateway connecting AS 1 to AS 2 may send the pair $\langle AS1, nid \rangle$ to AS 2. At that point, AS 2 will have discovered a route to network nid. AS 2, in turn, may advertise this information to its own neighbors, in which case it would send the tuple $\langle AS2, AS1, nid \rangle$ to its neighbors.

You may have noticed that this approach toward discovering routes is essentially the same as the one applied in the Bellman-Ford algorithm we discussed in Chapter 5. And indeed, the core of the so-called **Border Gateway Protocol** (**BGP**) which is deployed for discovering routes between autonomous systems is exactly this routing algorithm. However, instead of only reporting distances, BGP requires that an AS advertises the complete path it found to a destination. This information will allow a recipient to decide whether it will actually use that path for routing packets. Generally, a gateway will keep only information on its discovered shortest path to a network. Information on paths that are longer is simply discarded. What we are thus seeing is that (1) border gateways learn about shortest paths to networks in other autonomous systems, and (2) advertise this information to their neighbors, allowing each, in turn, to discover paths to those networks as well.

With over 25,000 autonomous systems, each having many networks to which packets must be routed, it is clear that the information that must be stored at a border gateway can be huge. In principle, each gateway is required to have an entry for *every* discovered network. Even with using many sophisticated techniques to combine routing information, a border gateway is currently required to store close to 300,000 entries. These entries are exclusively used to decide to which next AS an incoming packet is to be routed. In addition, every gateway stores information on well over 800,000 routes. In principle, those routes cover all paths between networks in the Internet.

With this in mind, it may now be clear how we can discover the AS topology of the Internet: we simply retrieve the routing tables from border gateways in order to collect as many routes as possible. Of course, this is much easier said than done. As explained by Huston [2006], many ASes use multiple AS numbers resulting in approximately twice as many observed ASes as there are in reality. In addition, an AS may decide not to advertise a link to one of its neighbors because it simply doesn't want to support traffic of other ASes over that link. In other words, there may be a connection between two border gateways from different ASes, but this is not reflected in BGP routing tables. Another source of errors is the dynamicity of the tables: when a link is temporarily out-of-order, it may not show up in routing tables.

As an aside, the discovery of the AS topology brings up an important scientific question, namely to what extent does our input data accurately represent what we are trying to model. We will return to this issue when we discuss how to construct a graph model of the Web.

A snapshot of the AS topology

Essentially using the method just described, Chi et al. [2008] have collected data on how autonomous systems link to each other. Taking a single snapshot from October 2008, we obtain a network consisting of over 30,000 vertices and more than 100,000 edges. Figure 8.4 illustrates that we are apparently dealing with a scale-free network, although the data points do not quite fit a straight line.

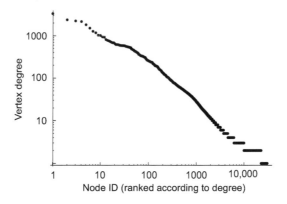

Figure 8.4: The degree distribution of the AS topology using BGP router data. The x and y axis are scaled logarithmically.

There are a number of interesting points to observe about this topology. First, it may be somewhat surprising to see how well connected some of the autonomous systems are. If we consider the degrees of the top-10 ASes, we find the following:

Rank:	1	2	3	4	5	6	7	8	9	10
Degree:	3309	2371	2232	2162	1816	1512	1273	1180	1029	1012

Not only do we see that the top AS is connected to more than 10% of all other ASes, we can also observe that this type of connectedness drops rapidly as one would expect from a scale-free network. As we discussed before, such a degree distribution may have a serious adverse effect on the robustness of the network, in the sense that a targeted attack by which we remove well-connected nodes may easily lead to partitioning the network.

Haddadi et al. [2008] have analyzed other properties of the AS topology found from BGP routers. Not only did they find high clustering coefficients for the top 1000 nodes, these nodes are also connected to each other forming an almost complete graph. In line with these observations is the distribution of shortest paths: most paths are no longer than three or four hops,

8.2. PEER-TO-PEER OVERLAY NETWORKS

and virtually all ASes are separated by a shortest path of maximum length six. Again, we see the small-world phenomenon occur in the network of autonomous systems.

> **Note 8.1** (More information)
> Unfortunately, just taking a snapshot of the AS topology may not provide enough information of what is going on. There are two problems that need to be addressed. The first one is caused by the fact that even with data from a large number of BGP routers, one can never be sure to have captured all existing peering relationships between autonomous systems. In fact, it turns out that finding the actual links at a given time may indeed be very difficult. The second problem has to do with the fact that large real-world networks are in continuous flux: links and nodes may appear to come and go all the time due to intermittent failures, making it more difficult to identify truly new peering relationships or those that have been discontinued.
>
> Consequently, when we're interested in identifying the real topology of the AS network, we need to do a bit more than just analyze a few snapshots. We will not go into further details here, but refer the interested reader to Chi et al. [2008] and Raz and Cohen [2006]. The latter provide evidence that more than 30% of the existing links are missing from the AS topologies derived from BGP routers. In fact, Oliveira et al. [2008] argue that only the observed links between the autonomous systems for tier-1 networks are reasonably accurate. For tier-3 networks, using BGP routing information is argued to be highly incomplete.

8.2 Peer-to-peer overlay networks

As will have become clear by now, the Internet is simply huge. In practice, we see that the Internet is used as a universal platform for a wide variety of applications. Perhaps the most well-known application is the Web, which we will discuss in Section 8.3. In many cases, Internet applications are organized according to what is known as a **client-server architecture**. In this case, the core of an application is hosted by a special computer, known as a **server**. The rest of the application consists of a program hosted on a so-called **client computer**. This client program can send a request to the server, where it is processed, after which the server sends a reply back to the client. A well-known example of this client-server architecture is actually the Web: the client program is formed by a Web browser; the server is the computer maintaining a specific Web site.

A client-server architecture can be represented by a simple graph in which clients and server are represented by vertices, and where each client vertex is joined with the vertex representing the server, as shown in Figure 8.5.

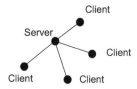

Figure 8.5: Representing a client-server architecture as a graph. In this example, there are four clients.

Although it would seem that the server in a client-server architecture may easily become a performance bottleneck, you need to realize that clients come and go quickly. In most cases, a client merely sends a request to the server, the server processes that request, and subsequently sends an answer back. After that, the client and server each go their own way. In graph-theoretical terms, the edge between a client and server will eventually be broken again. Nevertheless, in case we are dealing with requests that require substantial server processing time, or when responses require returning huge amounts of data, servers can indeed become a bottleneck because they can only process a limited number of requests per time unit. It is beyond the scope of this text to go into these matters in more detail. See Tanenbaum and van Steen [2007] for more information.

Since the late 1990s, researchers have been exploring alternative architectures to address scalability problems for large, distributed applications whose constituents are spread across the Internet. In principle, each constituent, called a **peer**, consists of a program that is being executed on a single computer. Each peer maintains a list, called a **partial view**, of other peers that form part of the distributed application. This partial view has the sole purpose to allow for the exchange of application-specific data between two peers. If we were to represent such a distributed application as a graph, each peer would be represented by a vertex and an edge would represent the fact that two peers would have each other in their respective partial views. Taking all these peers and their respective partial views into account leads to what is known as a **(peer-to-peer) overlay network**: a communication network between the constituents comprising a distributed application.

Structured overlay networks

One important type of overlay network is formed by networks that are organized in a structured fashion. In particular, the partial view of each peer is filled with references to very specific peers as opposed to having a partial view with references to randomly chosen peers. We will discuss the latter

The Chord peer-to-peer network

To make these matters concrete, let's consider the **Chord** peer-to-peer network [Stoica et al., 2003]. The principle behind Chord is relatively simple, which is also the reason why we'll use it to explain structured peer-to-peer networks. A survey of other, similar systems, is provided by Lua et al. [2005].

Chord is a distributed application that can be used to efficiently store and locate data across a huge collection of hosts. Each host is required to have a unique identifier, represented by an m-bit number. Typically, $m = 128$, meaning that there can be as much as $2^{128} \approx 3.4 \times 10^{38}$ identifiers. That's enough to fill every square *millimeter* land of the Earth with more than 2×10^{18} hosts. It should suffice for a while. In practice, this means that when a host needs to join a Chord network, it can simply generate its own random identifier without running any serious risk that some other host has generated the same identifier.

A host in a Chord network is assumed to store data. To keep matters simple, we assume that data is stored in a file, with each file having a unique key. Like host identifiers, each key is an m-bit number. The fundamental principle in Chord is that the file with key k is stored on the host with the smallest identifier id greater or equal to k. Computing if $id \geq k$ is done in modulo M arithmetic, where $M = 2^m$.

Note 8.2 (Mathematical language)
Recall that modulo M arithmetic is applied to integer numbers, mapping all numbers to values between 0 and $M - 1$. A common notation is k mod M. So, with $M = 32$, we would have:

k	k mod 32
4	4
31	31
32	0
−5	27
−31	1

To illustrate, consider a Chord network with $m = 5$, meaning that $M = 2^5 = 32$. Suppose we have peers (i.e., hosts) with identifiers 1, 4, 9, 11, 14, 18, 20, 21, and 28. It is convenient to represent this system as a ring, as shown in Figure 8.6. We simply denote the peer with identifier p as peer p. The actual peers are shown as gray-colored circles; the rest of the unused identifiers are represented by dashed circles. As shown, the peer with $id = 1$ will be responsible for storing files with key 29, 30, 31, 0, and 1, respectively. Indeed, in modulo M arithmetic we may have that $1 \geq 31$.

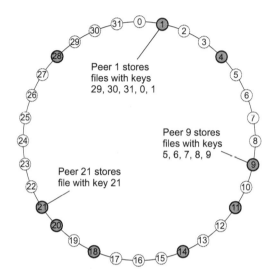

Figure 8.6: The representation of a Chord network as a ring.

The peer responsible for storing a file with key k is called the successor of k:

Definition 8.1: *Consider a file with key k. In a Chord peer-to-peer network, the peer with the smallest identifier $p \geq k$ is called the **successor** of k, denoted as $succ(k)$.*

Perhaps a bit confusing, but it is important to note that if $p = k$, $succ(k) = p$.

Central to the design of Chord is efficiently looking up data by means of keys. A naive way of doing a lookup is as follows. Assume that the peer with identifier p (i.e., peer p) is requested to look up a file with key k. If $p < k$, peer p can simply forward the request to its left-hand (i.e., clockwise) neighbor in the ring, a process which is repeated until the first peer is reached, say q, with $q \geq k$. Likewise, if $p > k$, peer p can still simply forward the request to its left-hand neighbor, until a peer q is found with the *smallest* identifier $q \geq k$. It is not difficult to see that this search strategy would, on average, require that a request is forwarded $\frac{1}{2}n$ times, where n is the total number of peers. If $n = 10,000$, it would take forever to locate the file.

A much more efficient approach is to let every peer store "shortcuts" to other peers at increasingly longer distances. These shortcuts are stored in

8.2. PEER-TO-PEER OVERLAY NETWORKS

a peer's partial view, which is called a **finger table** in Chord. Each finger table FT_p of peer p consists of m entries, numbered $1, 2, \ldots, m$, and denoted as $FT_p[1], \ldots, FT_p[m]$. Entry i contains the successor of key $p + 2^{i-1}$:

$$FT_p[i] = succ(p + 2^{i-1}).$$

In other words, entry i contains a shortcut to the peer responsible for key $p + 2^{i-1}$. The finger tables for our example Chord network from Figure 8.6 are shown in Figure 8.7.

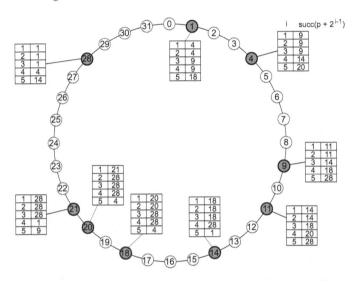

Figure 8.7: Finger tables for the peers from Figure 8.6.

Let's check a few of these finger tables:

- Consider $FT_4 = [9, 9, 9, 14, 20]$. $FT_4[1]$ should contain $succ(4 + 2^{1-1}) = succ(5)$. The peer responsible for key 5 is indeed 9. The same holds for $FT_4[2] = succ(4 + 2^{2-1}) = succ(6)$ and $FT_4[3] = succ(4 + 2^{3-1}) = succ(8)$. Likewise with $FT_4 = succ(4 + 2^{4-1}) = succ(12)$, the responsible peer for key 12 is indeed peer 14. Finally, $FT_5 = succ(4 + 2^{5-1}) = succ(20)$, which brings us to peer 20.

- For peer 21, we have $FT_{21} = [28, 28, 28, 1, 9]$. For the first three entries, we are seeking the successor peers for $21 + 1$, $21 + 2$, and $21 + 4$, respectively, which is indeed peer 28. $FT_{21}[4] = succ(21 + 8) = succ(29)$, for which peer 1 is responsible. Finally, $FT_{21}[5] = succ(21 + 16) = succ(37)$. Because we need to apply modulo 32 arithmetic, we find that $FT_{21}[5] = succ(37 \bmod 32) = succ(5)$, which leads us to peer 9.

It is now not hard to imagine how an arbitrary peer p receiving a request to look up key k proceeds: it looks in its finger table to identify peer q satisfying

$$q = FT_p[i] \leq k < FT_p[i+1]$$

In case $p < k < FT_p[1]$, q is selected to be $FT_p[1]$. Likewise, if $FT_p[m] \leq k$, q is selected to be $FT_p[m]$. The lookup request is then forwarded to q. This process is repeated until the request arrives at the peer responsible for k.

Key	Initial peer	Lookup path
15	4	$4 \to 14 \to 18$
22	4	$4 \to 20 \to 21 \to 28$
18	20	$20 \to 4 \to 14 \to 18$

Figure 8.8: Some example lookup paths.

To illustrate, consider the lookup requests from Figure 8.8. Using the notation $k@p$ to denote that a request for key k is initially issued at peer p, we have:

15@4: Because $FT_4[4] \leq 15 < FT_4[5]$, the request is forwarded from peer 4 to peer $FT_4[4] = 14$. There, we need to apply the rule that $p = 14 < 15 < FT_p[1]$, so that the request is forwarded to $FT_{14}[1] = 18$, where it reaches its destination.

22@4: For this request, we find that $FT_4[5] \leq 22$, so that the request is forwarded to peer $FT_4[5] = 20$. There, key 22 satisfies $FT_{20}[1] \leq 22 < FT_{20}[2]$, so that it is forwarded to peer $FT_{20}[1] = 21$. Again, noting that $p = 21 < 22 < FT_p[1]$, the request reaches its destination peer 28.

18@20: This is a somewhat tricky case. First, note that because $p = 20 \not< 18$, we cannot forward the request to $FT_p[1]$. Instead, we eventually find that $FT_{20}[5] < 18$, so that it is forwarded to peer 4. From there, it is easy to see that by using the same reasoning as for request 15@4, we find the remaining path $4 \to 14 \to 18$.

The Chord graph

Now that we have explained the basic principles of Chord, let's consider it from a different point of view, namely as graph. It should be clear how we can represent a Chord network as a (directed) graph: each peer is represented as a vertex and if peer p has a reference to peer q in its finger table, we add the arc $\langle \overrightarrow{p,q} \rangle$. This leads to the graph representation of our example

8.2. PEER-TO-PEER OVERLAY NETWORKS

Chord network shown in Figure 8.9. Note that we should indeed represent a Chord network as a *directed* graph: if p has q in its finger table, this does not mean that q also has a reference to p.

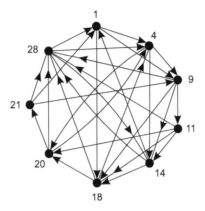

Figure 8.9: The representation of the Chord network from Figure 8.7 as a directed graph.

Of course, Chord networks become interesting when we consider ones with many peers. Already by looking at a network with 100 peers as shown in Figure 8.10, we can observe that we may be dealing with a small-world graph. First, we see that every vertex is joined with a vertex opposite its own position in the ring. In particular, suppose we would renumber the n vertices to $1, 2, \ldots, n$ and again use the distance metric

$$d_2^n(i,j) \stackrel{\text{def}}{=} \min\{|i-j|, n-|i-j|\}$$

which we introduced in Chapter 7 in the proof of Theorem 7.3. Recall that this metric measured the distance between vertices along the "outer ring" of the graph. With d_2^n, we then see that (virtually) every vertex p is joined with a vertex at roughly distances $\frac{1}{2}n, \frac{1}{4}n, \frac{1}{8}n, \ldots, 1$.

This observation also suggests that the average path length, which corresponds to the average number of vertices to which a lookup request needs to be forwarded, will most likely be proportional to $\log_2(n)$. To test this hypothesis, we can generate a series of Chord networks and compute the average path length for each of them. Figure 8.11 shows the result for a series of such networks. The figure also shows a logarithmic function that can be found using a standard curve-fitting method (see again Note 6.1 on page 136). As can be observed, not only does the average path length increase logarithmically with the size of the network, it is also relatively small. Note that the path length has been computed for a *directed* graph.

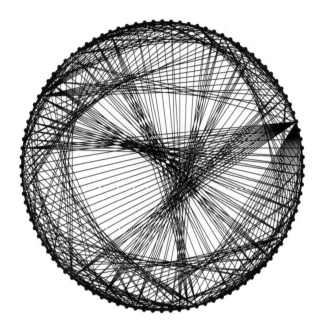

Figure 8.10: A Chord network with $m = 28$ and $n = 100$ peers (orientation not shown).

Note 8.3 (More information)
That the average path length indeed increases only logarithmically can be proven formally as follows (see also Stoica et al. [2003]).

Theorem 8.1: *Consider a Chord network with m-bit identifiers and n peers. The number of peers that need to be contacted in order to look up a key k is proportional to* $\log_2(n)$.

Proof. Assume that we issue a lookup request for key k at peer p. Let z be the peer that immediately precedes $succ(k)$. Assuming that $p \neq z$, peer p will forward the request to the closest predecessor of k that p can find in its finger table. This is exactly the rule stating that the next peer q should satisfy:

$$q = FT_p[i] \leq k < FT_p[i+1]$$

Let i be such that z is in the interval $[p + 2^{i-1}, p + 2^i)$. What p will do is contact the first peer q in this interval, which is precisely $succ(p + 2^{i-1})$. Note that $|q - p| > 2^{i-1}$, but at the same time $|q - z| \leq 2^i - 2^{i-1} = 2^{i-1}$. In other words, q lies numerically closer to z than to p, i.e., $d_2^M(p,q) > d_2^M(q,z)$ where $M = 2^m$. This also means that $d_2^M(q,z) < \frac{1}{2} d_2^M(p,z)$. The latter observation is important,

8.2. PEER-TO-PEER OVERLAY NETWORKS

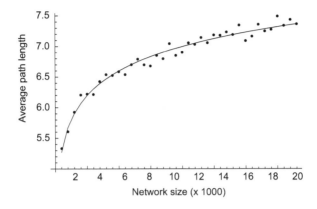

Figure 8.11: The average path length for a series of Chord networks with $m = 28$ and increasing number peers.

---

for it means that each time a request is forwarded, the distance to z measured according to metric d_2^M is at least halved.

What does this mean after having forwarded the request $2\log_2(n)$ times and reaching, say, peer r? Considering that we half the distance to z in every step, we will have a total reduction of $(\frac{1}{2})^{2\log_2(n)} = 2^{-2\log_2(n)} = 2^{\log_2(n^{-2})} = 1/n^2$. The distance between p and k will be at most 2^m, meaning that in $2\log_2(n)$ steps, the distance between r and z will be at most $2^m/n^2$. Because we are assuming that peer identifiers and keys are drawn uniformly at random, the probability that we have chosen a peer identifier from an interval of length L for an n-peer Chord network, is equal to $n \times L/2^m$. In other words, the probability that there is peer with an identifier between k and r, is equal to $n \times (2^m/n^2)/2^m = 1/n$, which is negligible for large n. We conclude that the number of peers that need to be contacted before resolving a lookup request is proportional to $\log_2(n)$. □

Let us take a look at some other properties of Chord networks, starting with the degree distribution. Because we are dealing with a directed graph, we should make a distinction between the distribution indegrees and outdegrees. Consider a Chord network with $n = 10000$ peers and using m-bit identifiers. Figure 8.12 shows the histograms for the indegrees as well as the outdegrees. When it comes to the indegrees, the distribution seems to follow an exponential curve (note, however, that we are not dealing with a power-law distribution). This also means that there are a few peers with many incoming arcs, in turn, meaning that they may need to process many lookup requests. The outdegrees are more or less symmetrically centered around 13.5. As argued by Stoica et al. [2003], each finger table will have only approximately $\log_2(n)$ unique entries, which in our example comes down to $\log_2(10,000) = 13.3$.

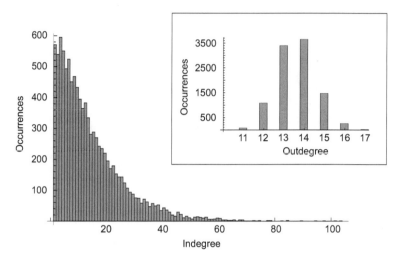

Figure 8.12: The distributions of indegrees and outdegrees for a Chord network with $n = 10000$ peers using 28-bit identifiers.

What about the clustering coefficient? To keep matters simple, we drop the orientation of a Chord network and compute the clustering coefficient of the corresponding undirected graph for various network sizes. The result is shown in Figure 8.13. First, compared to an Erdös-Rényi random graph, we see that the clustering coefficient is very high. Moreover, the clustering coefficient only slowly decreases when the network grows. Combined with the fact that the average path length is low, we may indeed conclude that Chord networks constitute small-world networks.

Random overlay networks

Processes in distributed applications such as Chord apply strict rules for maintaining partial views, effectively leading to a well-structured overlay. In contrast, in the case of random overlay networks, also referred to as **unstructured peer-to-peer networks**, the goal is keep a high degree of randomness in the partial view. In other words, the goal is to let entries refer to seemingly randomly chosen peers. In this section, we will take a closer look at a class of random overlay networks that are constructed through what is known as **gossiping**.

8.2. PEER-TO-PEER OVERLAY NETWORKS

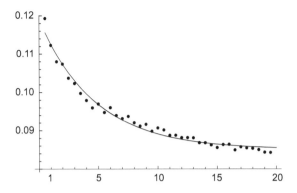

Figure 8.13: The clustering coefficient Chord networks of various sizes using 28-bit identifiers.

A framework for epidemic-based networks

As said, many unstructured peer-to-peer networks maintain an overlay that resembles a random graph. There are numerous ways to do this, and in many cases we see that this maintenance is done using centralized components. In other words, special central servers are used assist in maintaining some form of randomness in the overlay network. A fully decentralized approach can be achieved by making use of what are known as **epidemic protocols**. In an epidemic protocol, a peer (again, meaning a host) uniformly at random chooses another peer to exchange data with. It's as simple as that.

More formally, we have the following. Consider a collection of peers $P = \{p_1, p_2, \ldots, p_n\}$, each capable of storing a potentially very large collection of files. Each file f has a version number $v(f)$ telling how often the file has changed. To keep matters simple, we assume that each file has exactly one associated peer $own(f)$ that is allowed to change that file. Let $v(f, p)$ denote the version of file f currently stored at peer p, and $FS(p)$ the set of files stored at p. If f is not stored at peer p, then $v(f, p) = 0$. It should be obvious that

$$\forall f, p : v(f, own(f)) \geq v(f, p)$$

The principal goal of an epidemic protocol is to make sure that every update to a file is disseminated to all peers. To this end, each peer $p \in P$ *periodically* chooses uniformly at random another peer $q \in P$, and proceeds as follows[2]:

1. for all $f \in FS(p)$: if $v(f, p) > v(f, q)$, then $FS(q) \leftarrow FS(q) \cup \{f@p\}$, possibly replacing an older version of f that was stored at q.

---

[2] We use the notation "$f@p$" to denote the file f as stored at peer p.

2. for all $f \in FS(q)$: if $v(f,p) < v(f,q)$, then $FS(p) \leftarrow FS(p) \cup \{f@q\}$, again possibly replacing an older version of f that was stored at p.

Note that after these two steps, both peers p and q have exactly the same set of files and for each file also the same version. This protocol forms the core of a scheme that was proposed to maintain replicated databases by Demers et al. [1987]. It is widely applied in modern distributed systems to efficiently disseminate information.

Note 8.4 (More information)
Epidemic protocols are extremely efficient when it comes to spreading data. To see why, consider a collection of n peers in which initially each peer except p_1 stores nothing. Peer p_1 stores a data item d. Let us first consider two simple strategies:

1. When peer p contacts q, only if p already stores d and q does not, will p send d to q. In other words, p *pushes* data item d to q if p has it stored, otherwise nothing happens.
2. When peer p contacts q, only if q already stores d and p does not, will q return d in response to p's request. In other words, p *pulls* data item d from q if p does not yet have it stored, otherwise nothing happens.

We assume that each peer contacts another peer once every T time units. T is called the **cycle time**: after T time units we know that every peer has contacted exactly one, randomly chosen other peer. We therefore also say that a cycle has completed after T time units. Let ρ_i^+ be the probability that an arbitrary peer has not yet obtained d after i cycles in the *push* case, and ρ_i^- the same probability but when we apply only a *pull* strategy.

For the pull case, it is not difficult to see that $\rho_{i+1}^- = (\rho_i^-)^2$: the peer did not yet have d in the ith cycle and it contacted another peer who also did not have d during the ith cycle.

The push case is only slightly more complicated. In order for a peer p to stay bereft of d, it will have to be contacted only by peers who also do not have d stored. In other words, none of the peers that had stored d during the ith cycle should contact p. The probability that one such peer does not contact p is $(1 - \frac{1}{n-1})$: it has $n-1$ peers to choose from, one of them being p. The probability that it *will* contact p is therefore $\frac{1}{n-1}$. If ρ_i^+ is the probability that a peer will not have seen d up until the ith cycle, we can expect that a fraction of $(1 - \rho_i^+)$ will have already stored d. In other words, we can expect a total of $n(1 - \rho_i)$ peers to have stored d after reaching the ith cycle. None of them should contact peer p if we want p to stay ignorant of d, leading us to:

$$\rho_{i+1}^+ = \rho_i^+ (1 - \frac{1}{n-1})^{n(1-\rho_i^+)}$$

8.2. PEER-TO-PEER OVERLAY NETWORKS

To get an impression of the speed by which a data item is spread across a network using only the push or pull approach, consider Figure 8.14. We consider a 100,000 node network in which initially only a single peer stores data item d. Clearly, both approaches show that as soon as the data item has been sufficiently spread (which happens around cycle 13), dissemination speeds up tremendously.

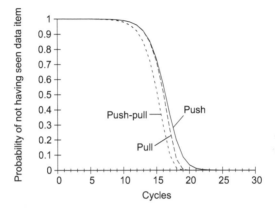

Figure 8.14: An illustration of the speed of epidemic-based dissemination of data.

Finally, when applying both pull and push when exchanging data, for a peer to remain ignorant of d, it should neither contact a peer that has stored d, nor be contacted by a peer storing d. In other words, $\rho_i^{+-} = \rho^+ \cdot \rho_i^-$. Again, we see that once dissemination has reached a few peers, within only a few cycles the whole network will know about d.

When dealing with very large networks, epidemic protocols bump into a practical problem: how can a peer uniformly at random select another peer? In principle, doing so requires that the selecting peer knows all the other peers in the network, yet having such complete knowledge is infeasible. Fortunately, we can take a much simpler approach by again considering partial views and letting peers exchange entries using an epidemic protocol. The crucial difference with a normal epidemic protocol is that a peer p now selects another peer chosen *from its partial view*. This is best explained by assuming that each peer is split into two programs that are executed simultaneously, called its active part and its passive part, respectively. The two programs are outlined in Figure 8.15.

Let us first concentrate on the active part of a peer p. We use the notation PV_p to denote the partial view of peer p. As shown, peer p waits for a fixed amount of time, after which it selects a peer q from its partial view PV_p,

Active part	Passive part

repeat
  wait T
  $q \leftarrow$ select 1 from PV_p
  $R_p \leftarrow$ select s from PV_p
  send $R_p \cup \{p\} \setminus \{q\}$ to q
  skip
  receive R_q^p from q
  $PV_p \leftarrow$ select m from $PV_p \cup R_q^p$
until forever
	repeat
	  skip
	  skip
	  skip
	  receive R_p^q from any p
	  $R_q \leftarrow$ select s from PV_q
	  send $R_q \cup \{q\} \setminus \{p\}$ to p
	  $PV_q \leftarrow$ select m from $PV_q \cup R_p^q$
	until forever

(a)	(b)

Figure 8.15: The basics of an epidemic exchange of references from partial views. Each peer consists of an (a) active part and a (b) passive part.

which it will later on exchange data with. This waiting time, or **cycle time** as it is called, is the same for every peer. We assume that within a cycle time, each peer will initiate an exchange with another peer *exactly once*, albeit that every peer does this at a different moment. When all peers have finished such an exchange, we say that a **round** has completed.

After peer p has selected q, it continues to select s entries from PV_p (we assume $s \geq 1$), denoted as the set R_p. This set, extended with a reference to p itself but always excluding q, is then sent to q. Meanwhile, peer q has been passively waiting for any incoming message. In our example, it receives a message from peer p, in particular the set R_p^q. Of course, we have that $R_p^q = R_p \cup \{p\}$. As with p, peer q will then select s entries from its own partial view, and send those along with a reference to itself back to p. At this point, both p and q are in the same state. Conceptually, each first adds the references received from its peer to its partial view, and then shrinks the partial view to a fixed size of m entries, bringing it back to the original size.

In our explanation, we have deliberately left open many choices. Indeed, Figure 8.15 can be considered as a framework for a wide variety of epidemic-based protocols, as discussed extensively in Jelasity et al. [2007]. For example, should p select a peer q *randomly* from its partial view, or perhaps the the peer that has been in its view the longest? Likewise, there are different choices for selecting the s references to be sent to q: random ones, the freshest ones, the oldest ones, etc. Finally, we need to decide on how to shrink the partial view again to its original size. In the following, we will concentrate one specific protocol that fits this framework, called Newscast.

8.2. PEER-TO-PEER OVERLAY NETWORKS

Newscast: an epidemic-based peer-to-peer network

Newscast is an epidemic-based network, originally developed to facilitate large-scale computing on the Internet (see Jelasity et al. [2010] for an updated original description of Newscast). The protocol is extremely simple, yet shows interesting emergent behavior. We will discuss a slightly simplified version of Newscast, for which we return to the framework shown in Figure 8.15. In particular, we have the various parameters set as described in Figure 8.16.

Issue	Policy	Description
view size	$m = 30$	Each partial view has size 30
peer selection	random	Each peer uniformly at random selects a peer from its partial view
reference selection	random	A random selection of s peers is selected from a partial view to be exchanged with the selected peer
view size reduction	random	If the view size has grown beyond m, a random selection of references is removed to bring it back to size m

Figure 8.16: Parameter settings for the (adapted) Newscast protocol.

Let us first see whether Newscast is indeed capable of producing an overlay network that resembles a random graph. To start with, we consider the situation that every partial view is initially filled with references to randomly chosen peers, and then see how the protocol affects the degree distribution. As in the case of Chord, representing a Newscast network is done by modeling every peer as a vertex and a reference to peer q as stored in the partial view of peer p as an arc from p to q. For Newscast, the outdegree of every vertex is equal to m, so let's consider the indegree distribution. We consider a 10,000-node network. Figure 8.17 shows the distribution for the initial network and one after 200 **rounds**. As said, a round is defined as the situation in which each peer has initiated an exchange with exactly one other peer. In terms of Figure 8.15(a), a round corresponds to one iteration of the **repeat ... until** loop.

What can be clearly seen from Figure 8.17 is that the degree distribution changes from being symmetric to fairly skewed, with some peers having a relatively high indegree. When giving the matter some thought, this should actually come as no surprise: there is simply a nonzero probability that certain references to peers are spread across many peers because they are simply not removed when shrinking a partial view back to its original size m. By applying other strategies than just randomly selecting peers as we did

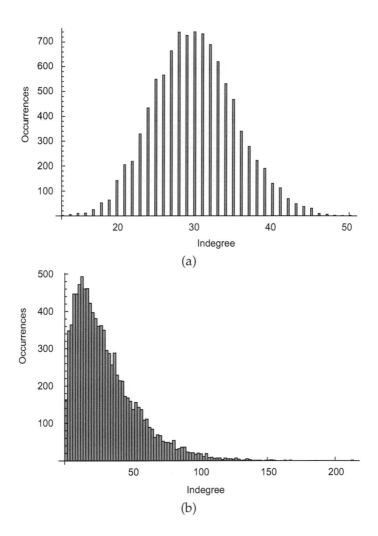

Figure 8.17: (a) The initial indegree distribution of a Newscast network, and (b) the situation after 200 rounds.

8.2. PEER-TO-PEER OVERLAY NETWORKS

for Newscast, we can achieve much better distributions (see, e.g., Voulgaris et al. [2005]).

Let's take a closer look at the average path length for Newscast networks. As with Chord, we take the orientation of the graphs into account, i.e., we consider the length of paths in the associated directed graph. The first thing to note is that Newscast networks are not always strongly connected. In other words, there are peers who cannot be reached by any other peer in the network. However, when conducting a reachability analysis, it turns out that such peers are few, and completely isolated. We therefore ignore them and concentrate only on the largest strongly connected component, which contains virtually all peers.

Figure 8.18 shows the average path length as the size of the network increases. In comparison to Chord, we see that the average path length is considerably smaller (see Figure 8.11). The figure also show the average path length for a comparable *directed* ER random graph. As can be seen, Newscast comes close to what we would expect to see from ER graphs when considering path lengths.

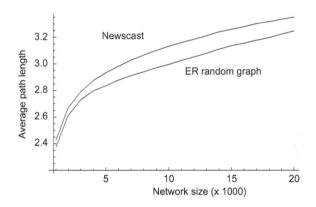

Figure 8.18: The average path length for Newscast networks of increasing size.

When it comes to the clustering coefficient, we see the following. Again, we simplify matters by dropping the orientation in the Newscast graph and consider its undirected counterpart. Figure 8.19 shows how the clustering coefficient evolves as the number of peers increases. For comparison, the figure also shows the clustering coefficient for an $ER(n,p)$ random graph, where p is taken equal to $30/(n-1)$. Recall that for an $ER(n,p)$ graph the average vertex degree is equal to $p(n-1)$. For our Newscast graphs, we have fixed the outdegree to 30, and thus also the average indegree. As a consequence, for a comparable $ER(n,p)$ graph we'll have $p(n-1) = 30$. In

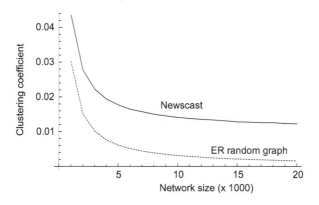

Figure 8.19: The evolution of the clustering coefficient in Newscast graphs as the number of peers increases.

contrast to Chord, we can see that Newscast appears to be much closer to Erdös-Rényi random graphs than small-world networks.

8.3 The World Wide Web

The Internet is the network that facilitates the undisputable biggest success of information systems: the World Wide Web, or simply **the Web**. Started around the late 1980s as a system to allow end users to easily browse through documents by means of **hyperlinks**, it has grown into a gigantic distributed information system with a virtually uncountable number of documents. Moreover, the system is in continuous flux: not only is content added and changed every minute, the number of participating sites that act as sources of information continues to grow at an exponential pace.

In this section we will explore the Web from the perspective of graphs. To do so, we first take a look at the basic organization that is needed to understand how its structure can be analyzed.

The organization of the Web

The Web is essentially organized into a vast number of **Web sites**. A site is a logical collection of Web documents with a uniquely associated **domain name**, such as, for example, www.distributed-systems.net. Using a site's domain name it becomes possible to access its documents as we explain below. NetCraft Ltd. reported the existence of close to 75 million active Web sites in the Fall of 2008. In the Summer of 2008, Google Inc. reported that they had discovered 1 trillion (i.e., 10^{12}) Web pages! When realizing that many

8.3. THE WORLD WIDE WEB

documents are not being seen by Google, imagining the actual size of the Web is virtually impossible.

A site, in turn, is hosted by a **Web server** or collection of servers. For our purposes, a Web server can best be thought of as a computer that is used to return a page. The browser that issued the request, is known as a **Web client**. The 75 million sites that were discovered by NetCraft were hosted on 65 million servers. Each such server essentially operates as shown in Figure 8.20. A browser issues a so-called **HTTP request** for a Web page to the server. HTTP stands for the **HyperText Transfer Protocol**, the standard communication protocol used in the Web. Such a request contains the domain name of the site, which is uniquely associated with the IP address of the server hosting the site. Before an HTTP request can be sent, the Web client first looks up the site's IP address using its domain name. This can be done using what is known as the **Domain Name System**, or simply **DNS**, but which we shall not discuss any further here (see, e.g., Albitz and Liu [2001] or Levien [2005] for further details).

The request as sent to the server contains an exact reference to the required document (which we describe shortly). The reference is subsequently processed by the Web server, allowing it to fetch the document from its local file system or database. At that point, the document is returned to the client.

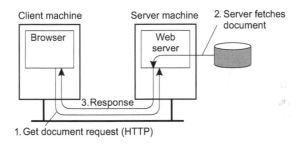

Figure 8.20: The basic communication between a Web client and server.

A document, that is, a Web page, may contain a reference to another document by means of a **hyperlink**. A hyperlink takes the form of what is known as a **Uniform Resource Locator**, or simply **URL**. To illustrate, consider the following URL:

http://www.distributed-systems.net/main.html

In this case, we have a reference to a Web page stored as the file main.html on the Web site with domain name www.distributed-systems.net. The additional "http://" tells us that this page can be accessed, or better, retrieved, by sending an HTTP request.

Once a client has received a page, it can fetch other pages through these URLs. For our purposes, it is important to realize that the combination of Web pages and the URLs they contain form the essential ingredients for constructing a graph. In particular, if we represent a page as a vertex, then clearly every URL contained in that page can be represented as an arc in a directed graph from that page to the page referenced by the URL. Summarizing, we are dealing with a graph estimated to consist of at least a trillion vertices and many more arcs.

Measuring the topology of the Web

Retrieving this so-called **Web graph** is practically undoable, if alone for the fact that it changes even more quickly than AS peering relationships. Unfortunately, there are several other problems that stand in the way of accurately measuring how pages link to each other. In this section we will go into further details on how the structure of the Web graph can be discovered.

Crawling the Web

In the beginning of the Web, documents were formatted using a relatively simple **markup language**: the **HyperText Markup Language** (HTML). A markup language is nothing but a series of commands that are inserted in the main text to tell a browser how it should render pages. For example, a command such as "<em>" can be used to *emphasize* a piece of text on a display. Most important for our purposes, is that a Web page can contain a reference to another page, such as:

    <a href="http://www.distributed-systems.net/main.html/">main page</a>

which tells a browser that if that reference is activated (e.g., by clicking with a mouse pointer on the text "main page" shown on the display), that it should fetch the page named www.distributed-systems.net/main.html. Life would be so much simpler if all references would be so explicit as in this example. Unfortunately, discovering how Web pages are linked to each other turns out to be a bit more complicated. To understand why this is the case, we need to delve into how the Web structure is actually measured.

A crucial tool for discovering Web structure is a so-called **crawler**: a program that automatically fetches pages that are referenced from a given page. The basic principle of a crawler is shown in Figure 8.21. Starting from a set of seed pages, it processes a page by extracting the references to other pages. Each of these references is appended to a list, called the **frontier**, reflecting the pages that have been found but not yet inspected. When a page has been processed, it is stored in a local repository.

8.3. THE WORLD WIDE WEB

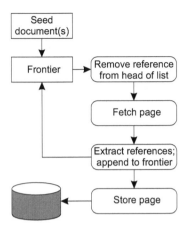

Figure 8.21: The principal operation of a Web crawler.

After having processed the seed pages, the crawler removes the reference that is at the head of the frontier and fetches the referenced page. It then simply extracts the references again, appending each of them to the frontier, after which the page is stored locally. It should be clear that in this way, one should indeed be able to fetch and store all pages that are *reachable* from the seed pages. That the repository for crawling and searching needs to be huge is exemplified by Google's approach. It has been estimated that by 2006, Google used approximately 500,000 servers, spread across the Internet (see also Barroso et al. [2003]). However, if we are interested only in discovering the topology of the Web, pages obviously need not be stored. In that case, we need "merely" build up a directed graph in which each vertex represents a fetched page, and every reference is represented by an arc.

As explained by Thelwall [2004] and Liu [2007], there are several difficulties that need to be dealt with. First, modern Web pages are no longer simple documents formatted in HTML. Instead, they may consist of different parts, some of which are complete programs (written in, for example, JavaScript). Finding references in such documents can be close to impossible, certainly if their creators have deliberately applied techniques to obfuscate references. Obscuring references is sometimes done on purpose to prevent Web pages from being indexed.

Second, many Web pages nowadays are not stored statically in file systems at a server's site, but are instead constructed and composed dynamically from a database query that is effectively part of the HTTP request. The problem is aggravated when the server is using programs to completely generate pages to be returned to the requesting client. As a consequence, we see that many references in the returned page are often personalized (i.e., based on specific information associated with the client), but also that the same request may return different pages (i.e., pages are also dependent on when they were requested). Conceptually, this means that the graph that represents the Web of pages that refer to each other, changes not only because edges are different all the time, but also because vertices effectively often exist only once and then disappear again for good.

Thirdly, and related to dynamic Web pages, crawlers need to be aware of **spider traps**. In this case, the references returned to a crawler depend on the order in which the crawler has visited pages from a given site. It may thus happen that when a crawler has fetched page A and discovered a reference to page B, that the server hosting B may generate a reference r_A to page A again that is contained in B, but that is interpreted by the crawler as a *new* reference (i.e., it fails to recognize that r_A refers to A, which it had already analyzed).

Finally, Web sites may simply install special files that are required to be read by all crawlers and which specify exactly which parts of the Web site are not to be inspected by crawlers. Although there is nothing that prevents a crawler to still inspect those parts, when such behavior is discovered, an administrator will most likely prevent any traffic from the site from which the crawler is operating.

Sampling the Web topology

There are other issues that make Web page discovery difficult, but one in particular is important when focusing on discovery topologies. It will come as no surprise that being able to fetch *all* Web pages, and thus building an accurate Web graph is practically impossible. By the end of 2008, the number of Web pages that have been discovered and indexed by search engines (also referred to as the **surface Web**), is estimated to be approximately 25 billion (i.e., 25×10^9). The actual size of the Web is likely an order of magnitude larger. Therefore, to get an impression of any network statistics regarding the Web graph, we are forced to consider only a *sample*. In other words, to discover certain properties of the Web graph we necessarily need to resort to collecting a subgraph. The question is how to make sure that such a subgraph is representative for the structure of the entire Web graph.

To this end, Becchetti et al. [2006] made a comparison between several crawling strategies. Note that when a crawler collects pages, it appends the references it finds to the frontier. This opens up several alternatives for inspecting next pages. In Figure 8.21 we suggested that pages are fetched from the head of the frontier. This is one common strategy, which leads to what is known as a **breadth-first inspection**. What happens is that first

8.3. THE WORLD WIDE WEB

all seed pages are inspected. When this is completed, the crawler inspects the pages that are directly linked from the seed pages, that is, at distance 1. Subsequently, the pages at distance 2 from the seed pages are inspected, and so on.

An alternative approach is not to select the head of the frontier, but to randomly select a reference from the frontier each time a new page is to be inspected. Also, one can take the popularity of a page into account, for example by considering the number of pages that are know to point to it (i.e., the *indegree* of a page). This latter strategy is closely related to the strategy followed by Google to determine the importance of a Web page, known as **PageRank** [Brin and Page, 1998] .

An important conclusion from their study, is that breadth-first inspection of pages leads to reasonable subgraphs, provided that these graphs by themselves are relatively large. For many of their network statistics, it turned out that a subgraph had to contain approximately 50% of the original set of vertices in order to produce representative results. This is actually quite a dramatic result, as it seems to imply that obtaining a representative sample of the Web may turn out to be extremely difficult.

And indeed, a recent study by Serrano et al. [2007] shows that there may be significant differences between various samples. Before we go into details, let us first consider some important structural properties of a **Web subgraph**. By the latter, we mean a graph that has been obtained by crawling a substantial number of Web pages and subsequently representing the pages and links between them as a directed graph.

In their famous study of two crawls of the AltaVista search engine comprising a set of over 200 million pages and 1.5 billion links, Broder et al. [2000] suggested to represent the Web as the **bowtie** shown in Figure 8.22. An interesting aspect of their study was that their sample most likely covered close to 16% of the surface Web at that time, which may be argued to be large enough to be considered representative.

Broder et al. made a distinction between the following groups of Web pages:

SCC The Strongly Connected Component (**SCC**) consists of a group of Web pages of which the corresponding directed graph is strongly connected. In other words, between any pair of vertices there exists a directed path from one vertex to the other.

IN This group of **IN** pages cannot be reached from any page in the SCC, but the SCC can be reached from pages in IN. More formally, for every vertex $v \in$ IN and $w \in$ SCC, there exists a directed (v, w)-path but no directed (w, v)-path.

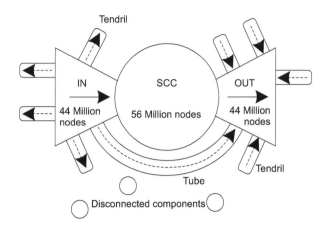

Figure 8.22: The macroscopic structure of the Web [Broder et al., 2000].

OUT Pages in **OUT** can be reached from the SCC, but are not part of the SCC. In particular, this means that for any vertex $v \in$ OUT and $w \in$ SCC, there exists a directed (w, v)-path, but no (v, w)-path.

TENDRILS A **tendril** is a collection of pages connected to either IN or OUT, but whose pages do not belong to either IN, OUT, or SCC. For example, a tendril TEN connected to IN consists of pages that can be reached from one or more pages in IN, but any path from a page $v \in$ IN to a page in TEN will never lead to a page in SCC. Note that a tendril itself may form a strongly connected component. Furthermore, it may very well be the case that certain tendrils can be reached from a page in IN, but also offer a path to a page in OUT, while none of the pages in that tendril belong to SCC. In this case, the tendril is called a **tube** .

DISCONNECTED This group consists of pages that cannot be reached from any of the other four groups. Typically, these pages are never found when crawling the Web. Alternatively, if a crawler starts from a disconnected page, it will never reach any page in IN, SCC, OUT, or a tendril.

Broder et al. found that there were approximately 44 million pages in IN, OUT, and all the tendrils. The SCC consisted of roughly 56 million pages, and a total of some close to 17 million pages were disconnected. If we were to consider this sample representative for the entire Web, it should be clear that any crawler can easily miss a substantial part of all available Web pages.

8.3. THE WORLD WIDE WEB

For example, when the collection of seeds is drawn from OUT, or any of the tendrils, it will be impossible to reach SCC.

Returning to Serrano et al. [2007], these authors have shown that the selection of seed pages is important when it comes to finding the pages that matter. In fact, it turns out that even when considering very large samples, the ratio of pages in IN, OUT and SCC may vary widely. To give an idea of what we're dealing with, Serrano et al. considered four different large samples, of which the characteristic properties are shown in Figure 8.23. In Figure 8.23(b) we visualize the relative differences between IN, SCC, and OUT, and compare it to the structure found earlier by Broder et al. The conclusion is clear: despite the fact that we may be sampling a very large part of the Web, it is difficult to conclude that the sample may be representative for the entire Web graph. Apparently, we have not yet found a valid technique for representative sampling (see also Cothey [2004]).

Component	Sample 1	Sample 2	Sample 3	Sample 4
SCC	56.46%	65.28%	85.87%	72.30%
IN	17.24%	1.69%	2.28%	0.03%
OUT	17.94%	31.88%	11.26%	27.64%
Other	8.36%	1.15%	0.59%	0.02%
Total size	80.57M	18.52M	49.30M	41.29M

(a)

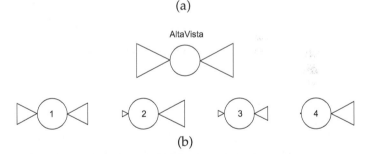

(b)

Figure 8.23: Comparing the relative sizes of IN, OUT, and SCC for different Web subgraphs. (a) The actual figures; (b) Relative comparison. From Serrano et al. [2007].

Characteristics of Web graphs

Let us now take a look at some of the properties of Web graphs. Various studies are based on the Stanford WebBase project [Cho et al., 2006], in which various crawls are being conducted and made available to the public.

Based on one such crawl, comprising more than 200 million pages, Donato et al. [2007] analyzed some of the characteristics of Web graphs.

As mentioned, Web graphs are directed: a hyperlink contained in page A referring to page B, is naturally represented by an arc from vertex A to B. In the case of vertex degree distributions, it is important to make a distinction between indegrees and outdegrees. Figure 8.24 shows the indegree distribution of the Donato et al. WebBase crawl after removing the nodes with very low indegree. In this case, as we have done before, the nodes have been ranked in descending order according to their indegree. The y-axis shows the *relative* indegree, with the highest indegree labeled as "1." We see that this curve again fits a power-law distribution quite reasonably, which has indeed been confirmed by Donato et al..

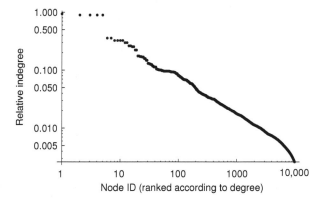

Figure 8.24: The distribution of indegrees of a WebBase crawl, as derived from [Donato et al., 2007].

It is interesting at this point to compare the actual indegree distribution with the PageRank algorithm that is used to distinguish important pages, i.e., pages that apparently contain much-wanted information. PageRank is used in Google and is based on indegrees. In particular, the *rank* of a page i is recursively defined as:

$$rank(i) = (1-d) + d \sum_{\langle \overrightarrow{j,i} \rangle \in E} \frac{rank(j)}{\delta_{out}(j)}$$

where $d \in [0,1)$ is known as a *damping factor*. What we see is that the rank of page i is determined by the page rank of the pages referring to i. Intuitively, this means that a page is considered important, not only if many other pages are referring to it, but notably when it is referred to by many other *important* pages. It is believed that for PageRank as used in Google, $d = 0.85$.

8.3. THE WORLD WIDE WEB

What the optimal value for d should be is unclear, but neither $d = 0$ or d close to 1 produces good ranks [Boldi et al., 2005]. As it turns out, there is only a weak correlation between the rank of a page and its indegree [Pandurangan et al., 2006]. In other words, it is not necessarily the case that a page with a high rank also has a high indegree, and *vice versa*. On the other hand, several studies show that if we compute the distribution of PageRank values, we again find a power-law distribution with scaling exponent 2.1. Again, we are confronted with the difficulty of drawing strong conclusions on the structure of the Web graph, even when using apparently reasonable metrics and sampling techniques.

For the outdegree distribution we observe a very different behavior, as shown in Figure 8.25. There is not a clear explanation why the outdegree does not fit a power-law distribution, but one possibility is that links *to* other pages need to be provided by the maintainers of Web pages. These maintainers may simply not have the patience (or the need) to include many hyperlinks in their pages.

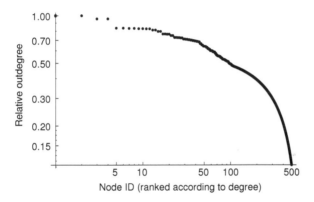

Figure 8.25: The distribution of outdegrees of a WebBase crawl, as derived from [Donato et al., 2007].

Let us now consider some other characteristics of Web graphs. In a study based on a simple Web crawl from 1998, Adamic [1999] constructed a graph by considering Web sites instead of pages. In particular, a graph was constructed by which vertex A has an arc to vertex B, if there was a Web page hosted by site A that referred to a page hosted by B. In this way, a graph was constructed comprising roughly 150,000 vertices (after discarding leaf vertices, i.e., having degree 1). For the underlying undirected graph, the average path length was estimated to be 3.1, while the clustering coefficient was found to be 0.1078. Clearly, we are dealing with a small-world network.

When considering the directed graph, the largest strongly connected component (SCC) consisted of approximately 65,000 sites, which is of the same order as the Web graph examined by Broder et al.. However, Adamic found an average shortest *directed* path length of 4.2, whereas Broder et al. found this to be equal to approximately 16. For the SCC of the latter, the average shortest path length in the underlying undirected graph was estimated to be 6.83. The difference between these observations may be caused by considering sites versus pages.

CHAPTER 9

SOCIAL NETWORKS

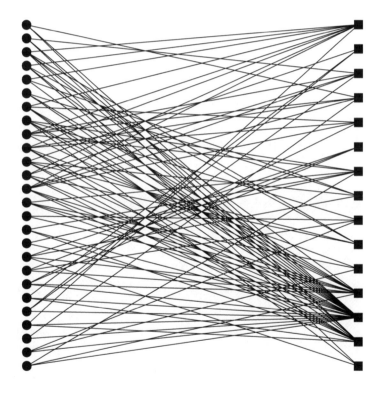

So far, our applications of graph theory have been taken from fairly technical communication networks. In these networks, the nodes are generally formed by computers or other devices. However, graph theory has also been extensively used to analyze social structures, also known as **social networks**. In a social network, a node represents a social entity, typically a person, an organization, and so on. An edge stands for a specific relationship between its incident nodes. In contrast to other areas in social sciences in which it is important to understand what characterizes social entities (e.g., by considering their attributes), social network analysis concentrates on the structure of relationships and tries to explain social phenomena from those structures. It should come as no surprise that graph theory plays a key role in social network analysis.

9.1 Social network analysis: introduction

Let us start our discussion with a motivating example to illustrate the applicability of social network analysis. We also briefly consider some historical background before delving into the specific metrics that are used to analyze social networks.

Examples

An illustrative example of how social network analysis can be effectively used is described in [Michael, 1997]. The example has also been used as a case study in de Nooy et al. [2005] from which we take the results of the analysis. The case is about a small wood-processing firm in which management proposed a new compensation package. This led to a strike, letting management believe that the communication to the workers had been far from optimal. They decided to have the social network analyzed. To this end, the workers were asked to indicate how often and with whom they discussed the strike. Frequency was measured on a 5-point scale, leading to a graph in which two people were linked if they frequently talked to each other. This graph is shown in Figure 9.1.

There are a number of properties that can be derived from this graph and which can be explained when we take a closer look at the individual members. First, there are apparently three clusters. The smallest one is formed by four workers, namely Eduardo, Domingo, Carlos, and Alejandro. These workers all used Spanish as their first language. Of these, Alejandro was most proficient in English. In addition, Bob spoke some Spanish, which most likely contributes to the link with Alejandro. Another cluster is formed by Frank, Gill, Ike, Mike, Bob, Hal, John, Lanny, and Karl (all represented as a gray-colored vertex). It turned out that these workers formed a group

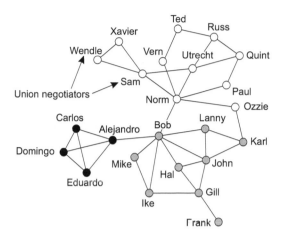

Figure 9.1: The relationship between workers on strike in a wood-processing firm.

of younger people, who did not speak that often with the older co-workers. The latter formed the third cluster, consisting of Norm, Ozzie, Paul, Sam, Wendle, Xavier, Vern, Ted, Utrecht, Russ, and Quint.

This clustering reflects what is known in sociology as **homophily**: the tendency of people to maintain stronger relationships with those who are similar to themselves.

The two union negotiators, Sam and Wendle, were initially responsible for proposing and opening the discussion on the new package. However, by taking a look at the network, it is not difficult to see that neither of them actually forms an ideal source for initiating communication. Intuitively, Bob and Norm, and to a certain extent also Alejandro, form the most important people in this network. And indeed, when management approached Bob and Norm directly to explain what the new package was all about, within only short time all workers understood the deal and were willing to negotiate. The strike ended.

Let us consider another example, this time concentrating on the Medici family. This highly influential and powerful family originated from Florence where Giovani di Bicci created the Medici Bank, making him one of the wealthiest men of Florence. His son, Cosimo de' Medici, continued along the same path as his father and is considered as the founder of the Medici dynasty, a dynasty which lasted for approximately 200 years. Cosimo de' Medici understood what it takes to get power and stay in power: make sure that the right people get married to each other. Padgett and Ansell [1993] analyzed the Medici dynasty during the first half the 1400s, including an overview of marriages between the Medici's and other fami-

9.1. SOCIAL NETWORK ANALYSIS: INTRODUCTION

lies, leading to the social network as shown in Figure 9.2.

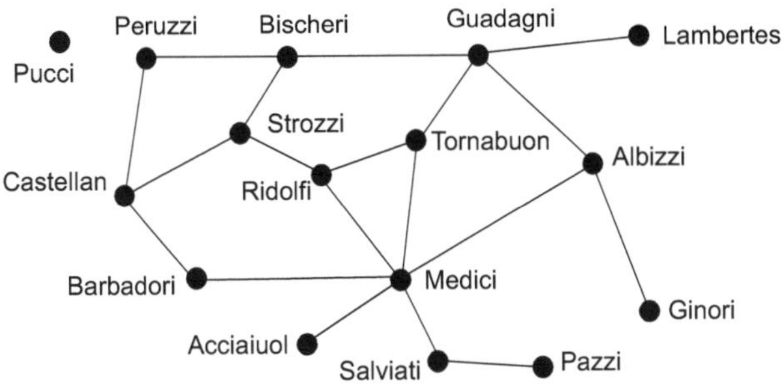

Figure 9.2: The relation between influential Florentine families in the beginning of the 15th century.

Following Jackson [2008] we provide a simple analysis of this network. A serious and in-depth analysis of the actual social relationships is given by Padgett and Ansell [1993]. For our analysis it is interesting to note that the Strozzi family not only had more money, but were also better represented in the local legislature. Nevertheless, the Medici's eventually became more powerful. Let's see what a possible reason could be, by looking at the **betweenness centrality**. Recall that the betweenness centrality $c_B(u)$ of a vertex u is defined as

$$c_B(u) = \sum_{x \neq y} \frac{|S(x,u,y)|}{|S(x,y)|}$$

where $S(x,u,y)$ is the collection of shortest (x,y) paths containing u, and $S(x,y)$ is the set of shortest paths between vertices x and y. If we normalize $c_B(u)$ by the possible pairs of families that u can connect, i.e., by $(n-1)(n-2)/2$, one can compute that the betweenness centrality for the Medici's is equal to 0.522, whereas this value is only 0.103 for the Strozzi's. Phrasing this differently, the Medici's were on more than 50% of all shortest paths in the network, whereas the Strozzi's covered only 10%. Indeed, when it comes to exerting power, the Medici's were seemingly in a much better position.

Historical background

Although social network analysis sometimes appears to be a novel discipline that recently emerged as another part of the science of networks, it is, in fact, since long a well-established area of research. Already in the beginning of the previous century, psychologists were using diagrams to represent relationships between social entities. An important contribution was made by Jacob Moreno who introduced the **sociogram** in the 1930s. In a sociogram, an individual is represented by a point, and relationships between individuals by lines – indeed, a graph. The importance of Moreno's sociograms lies in the fact that he suggested that one could derive specific characteristics from sociograms, like identifying influential people, identifying flows of information, and so on. And indeed, they have proven to be a powerful tool for discovering structure in social groups. We will return to one specific use below.

With Moreno's sociograms, the scene was set for further work in what is known as **sociometry**, which is all about quantitatively measuring social relationships. An important concept that arouse was that of a **triad**. A triad is a subgraph of a sociogram consisting of three points that *could be* connected to each other. Obviously, triads are related to triangles, which we discussed in Chapter 6. Formally, the distinction between a triad and a triangle is that in the latter the three vertices are joined with each other. For a triad, this need not be the case. Triads became important for studying the presence and evolution of social subgroups. For example, Cartwright and Harary [1956] developed a theory on **social balance** in which they considered subgroups of at least three individuals, as shown in Figure 9.3.

Figure 9.3: A triad to be analyzed for social balance.

In this particular case, the relationships between individuals was assumed to be symmetric: if Alice liked Bob, then Bob would also like Alice. If we represent "like each other" with a "+" and "dislike each other" with a "−," we can speak of balanced and imbalanced triads as reflected in Figure 9.4. The important observation here is that a sociogram is used to analyze a social group as a whole by considering all its members' perspectives on their relationships simultaneously. In other words, the focus is on discovering structures within the social group. In this way, one would be able to make statements about, for example, the stability or balance of an entire group, and to what extent one could expect that relationships would change (under the assumption that groups aim for balance). We will return to this phenomenon later in this chapter.

The idea of focusing on the discovery of global structures through the analysis of small-scale interactions, such as occurred in triads, led to new analysis techniques. In particular, researchers became interested in being able to identify different subgroups. In terms of graphs, this meant that

9.1. SOCIAL NETWORK ANALYSIS: INTRODUCTION

A–B	B–C	A–C	B/I	Description
+	+	+	B	Everyone likes each other
+	+	−	I	The dislike between A and C stresses the relation B has with either of them
+	−	+	I	The dislike between B and C stresses the relation A has with either of them
+	−	−	B	A and B like each other, and both dislike C
−	+	+	I	The dislike between A and B stresses the relation C has with either of them
−	+	−	B	B and C like each other, and both dislike A
−	−	+	B	A and C like each other, and both dislike B
−	−	−	I	Nobody likes each other

Figure 9.4: The possible balanced (**B**) or imbalanced (**I**) relations in a triad based on liking or disliking each other.

techniques needed to be developed that would allow the identification of components, yet allowing components to sometimes still be connected to each other. To illustrate, consider our example of the workers at the wood-processing firm again. Sociologists were interested to see which people actually formed groups within that community and were able to identify three of them, as mentioned before. These groups can be more easily visualized when considering the adjacency matrix of the associated network, as shown in Figure 9.5(a). For clarity, we omit the names of the workers. A cell (i, j) is colored black if worker i and j are linked to each other. By simply reordering the rows and columns, we obtain an equivalent matrix, shown in Figure 9.5(b). This last matrix reveals more strongly than the first one that there are indeed subgroups among the workers.

Although we have only visualized group boundaries, formal methods will indeed reveal that such groups can be identified. What we have shown in Figure 9.5 is known as **block modeling**, which was one of the earlier techniques for identifying subgroups. More techniques were eventually developed to allow for sometimes sophisticated clustering of nodes (see also Porter et al. [2009]).

It was not until the 1950s that researchers started talking more systematically about networks and would start using graph-theoretical concepts to express structural aspects of networks. The relationship between sociograms and the more rigorous approach implied by the use of mathematics was thus gradually introduced. However, it would take at least another decade until the ties between social networks and mathematics had come

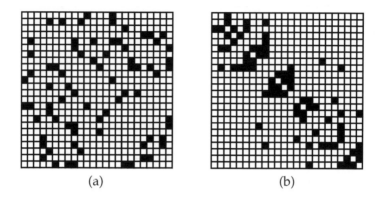

Figure 9.5: (a) The adjacency matrix of the network from Figure 9.1, and (b) the same matrix after reordering rows and columns. From [de Nooy et al., 2005].

to substantial strength. Of particular influence was the work by Mark Granovetter on what he called **weak ties**: links between different social clusters that proved to be essential for information dissemination, and thus reaching out to other groups than one's own [Granovetter, 1973]. Understanding Granovetter's work required a mathematical approach to social networks.

Social network analysis evolved steadily ever since then, and many rigorous techniques have been developed. We have now reached a new point. As mentioned, sociologists developed various models on how groups of people organize themselves. One particular famous one is the small-world organization, which we discussed in Chapter 7. The problem that researchers faced was how to validate those models: setting up sociological experiments with many participants is far from trivial as Milgram experienced in the late 1960s (recall that we discussed Milgram's experiments in Chapter 7). With online communities, researchers suddenly have tremendous sociological data sets in their hands. As we will also discuss in this chapter, we can apply similar analyses to these sets not only to validate models of how social networks evolve or how they are structured, but also to discover new properties that are inherently tied to the size of a network.

As argued by Kleinberg [2008], it is equally important that the analysis of these online social communities will perhaps put us in a much better position to devise large-scale distributed computer systems such as the fully decentralized peer-to-peer systems discussed in Chapter 8. We are already seeing better search strategies that are based on grouping peers by a notion of similarity, and many other phenomena related to social networking.

Sociograms in practice: a teacher's aid

Let us consider an example of a sociogram. One particular use of sociograms is in classrooms allowing a teacher to obtain better insight in the social structure of the class. In such cases, each child may be asked to list the three persons he or she likes the most (known as a positive nomination) or the least (i.e., negative nominations). An example is shown in Figure 9.6, which is based on material from Sherman [2000]. An entry (i, j) marked "+" indicates that child i liked child j, whereas a "−" indicates that i disliked j.

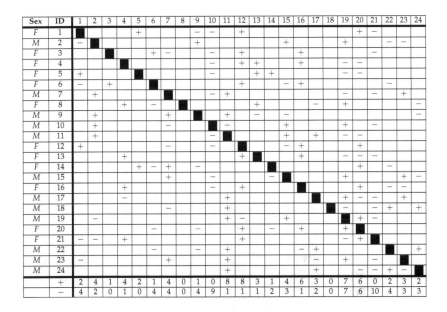

Figure 9.6: Data on the three most liked or disliked classmates.

When considering only the positive nominations, we obtain the social network shown in Figure 9.7(a). In this case, boys are represented by black-colored vertices whereas girls are shown as white-colored vertices. We instantly see that the two groups are more or less separated: boys and girls each tend to form their own subgroup, as is further illustrated after reordering the adjacency matrix, shown in Figure 9.7(b).

There are other issues that make this an interesting case. For example, by simply considering the distribution of indegrees, one can get an impression of the position of certain children. In this case, we should also consider the negative nominations as given by Figure 9.6. We see that children #11 and #12 are very popular (having very high indegrees for the positive

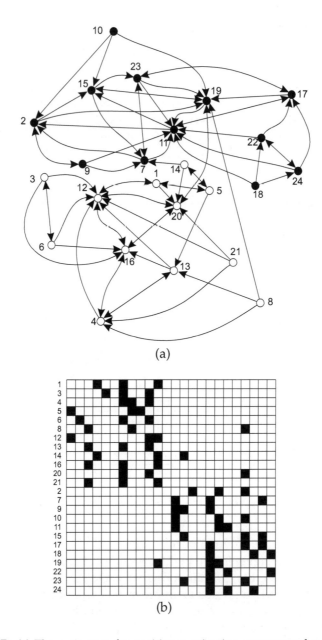

Figure 9.7: (a) The sociogram for positive nominations represented as a directed graph. Boys are represented by black-colored vertices; girls by white-colored vertices. (b) After reordering the adjacency matrix, the two subgroups become more apparent.

9.1. SOCIAL NETWORK ANALYSIS: INTRODUCTION

nominations), whereas #10 and #21 are very unpopular. There is much controversy regarding child #19 (and to a lesser extent #20), who received relatively many positive *and* negative nominations. There are also neglected children, namely those who are not mentioned at all (children #8 and #18).

Let us concentrate somewhat more on who is important and who is not by considering the largest strongly connected component of our classroom graph. This component consists of all children except #3, #6, #8, #10, #18, #21, #22, and #24. The eccentricity of a member was defined in Chapter 6 as the maximum distance of that member to any other member. For our subgroup, we obtain:

Child:	1	2	4	5	7	9	11	12
Eccentricity:	5	6	6	4	7	7	7	5
Child:	13	14	15	16	17	19	20	23
Eccentricity:	6	3	6	5	6	5	4	6

Interestingly, child #14 is closest to any other child, whereas the popular ones do not really differentiate from the others. When reconsidering Figure 9.7(a), we can see that child #14 is one of the few children who nominated a boy (#7) *and* a girl (#20). To see to what extent a child is close to every other member of the group, we compute the closeness values:

Child:	1	2	4	5	7	9	11	12
Close:	0.023	0.021	0.018	0.025	0.018	0.018	0.018	0.022
Child:	13	14	15	16	17	19	20	23
Close:	0.018	0.030	0.021	0.021	0.021	0.025	0.025	0.021

However, as we have argued before, closeness may not always be a good indicator of importance. For example, if child #14 was removed from the class, how harmful would that be for passing on information? In fact, it turns out that because #14 is really not that well connected, she also does not play a crucial role in these matters. Sociologists have introduced betweenness centrality as an indicator for importance. As explained before and in Chapter 6, this metric takes into account whether or not a vertex is lying on the shortest path between two other vertices. If we compute the betweenness centrality for each of our group members, we get the following values:

Child:	1	2	4	5	7	9	11	12
Betweenness:	0.140	0.153	0.050	0.105	0.083	0.007	0.155	0.220
Child:	13	14	15	16	17	19	20	23
Betweenness:	0.016	0.054	0.083	0.140	0.017	0.466	0.469	0.029

The results are interesting: without doubt children #19 and #20 play crucial roles when it comes to connecting the two groups of boys and girls, and thus in passing information between the two subgroups. Indeed, if we would remove either one from the subgroup, it would fall apart in the sense that we would no longer have a strongly connected component.

9.2 Some basic concepts

Now that we have given an overview of social networks and a typical example of how they can be applied, let's take a step further and consider a few of the more important concepts in social network analysis and how these concepts relate to the theoretical framework offered by graphs. In our discussion, we largely follow the structure as presented by Wasserman and Faust [1994].

Centrality and prestige

As we have mentioned, identifying important social entities forms a recurring topic in social network analysis. Up to this point we have introduced the following metrics to assist in finding those entities:

Vertex centrality: A metric that tells us to what extent a vertex is at the center of a graph, by considering its maximum distance to all other vertices. Typically, vertices "at the edge" of the network are generally considered less influential than those at its center.

Closeness: This metric considers the centrality as measured by the distance to each other vertex in the graph. The higher the value, the closer a vertex is to every other vertex.

Betweenness centrality: This important metric defines centrality of a vertex u by considering the fraction of shortest paths that cross u. The more such paths, the more important u is to be considered.

All of these metrics should be considered with care, as we illustrated in the previous section with our classroom example. For instance, we saw that a popular person may not be the one that is most efficient for spreading information.

Note further that these metrics can be defined for directed as well as undirected graphs, as they are all based on a notion of distance between vertices. However, when considering directed graphs, it is useful to make a distinction between the distance *to* other nodes (as one would use for measuring centrality), and the distance *from* other nodes. In particular, if we

9.2. SOME BASIC CONCEPTS

want to indicate the *prestige* of a vertex u, counting how many other vertices refer to u as a metric for prestige seems to make sense. In particular, we have:

Definition 9.1: *Let D be a directed graph. The **degree prestige** $p_{deg}(v)$ of a vertex $v \in V(D)$ is defined as its indegree $\delta_{in}(v)$.*

One can argue that degree prestige is a rather crude metric as it considers only direct relationships, namely the vertices that are adjacent to v. A more subtle way of measuring prestige is to also consider the vertices that can reach v through a directed path. In sociological terms, these vertices are called v's **influence domain**. In that case, we can compute the average distance to vertex v of the vertices in its influence domain, leading to the following definition.

Definition 9.2: *Let D be a directed graph with n vertices. The **influence domain** $R^-(v)$ is the set of vertices from where v can be reached through a directed path, that is, $R^-(v) \stackrel{def}{=} \{u \in V(D) |$ exists a (u,v)-path$\}$. The **proximity prestige** $p_{prox}(v)$ of a vertex v is defined as*

$$p_{prox}(v) \stackrel{def}{=} \frac{|R^-(v)|/(n-1)}{\sum_{u \in R^-(v)} d(u,v)/|R^-(v)|}$$

where $d(u,v)$ denotes the length of the shortest (u,v)-path in D.

Note that for proximity prestige we consider (1) the fraction of all vertices that can influence v (and exclude v), i.e., $|R^-(v)|/(n-1)$ and (2) the average distance of those vertices to v.

Note 9.1 (Mathematical language)
The definition of proximity prestige may not be instantly obvious, for which reason it is important to make sure that you understand what it means. The definition is also a good example to illustrate the precision of mathematics over a more verbal explanation.

First, it is important to realize why we are considering the *fraction* of influential vertices, i.e., $|R^-(v)|/(n-1)$. In doing so, proximity prestige can be expressed independent of the size of a graph, which is obviously an advantage as it allows us to more easily compare different networks. It should also be clear why we divide $|R^-(v)|$ by $n-1$ and not n: because we do not consider a vertex to be in its own influential domain, there are at most $n-1$ vertices who can.

Second, if we are going to consider the fraction of influential vertices, we should also consider the average distance of those vertices to v and not just merely the total distance. Again, this method of measurement allows us to better compare graphs.

> Finally, note that proximity prestige is always a value between 0 and 1. To this end, we first rewrite its definition to:
>
> $$p_{prox}(v) \stackrel{\text{def}}{=} \frac{|R^-(v)|^2/(n-1)}{\sum_{u \in R^-(v)} d(u,v)}$$
>
> so that we can more easily consider the case where there are no vertices in v's influential domain. In that case, $|R^-(v)| = 0$, and so is $p_{prox}(v)$. At the other end of the spectrum is the situation that we can reach v from every vertex, but moreover, each one is an in-neighbor of v. We then have that $|R^-(v)| = n-1$ and $\sum_{u \in R^-(v)} d(u,v) = n - 1$. As a consequence, we see that $p_{prox}(v) = 1$.

Let's reconsider our classroom example and take a look at proximity prestige within the largest strongly connected component. We make the following assumption: if child i has positively nominated child j, then the behavior of child j will affect child i. In other words, the directed graph of positive nominations can be seen as a directed graph of who influences whom by simply reversing the orientation of each arc. Using this reversed orientation, Figure 9.8 shows the distance between pairs of vertices, i.e., a cell (i,j) gives the shortest distance *from* vertex j *to* vertex i. These distances have been computed using the directed graph obtained by reversing the orientation of the graph from Figure 9.7.

The various values for proximity prestige lie quite close to each other, but again we see that children #19 and #20 have the highest score. Considering that these two also had the highest betweenness centrality, the social picture is becoming consistently clear.

One of the problems that social scientists have been struggling with is that the metrics we have been discussing so far consider importance without taking into account the importance of the nominating vertex. In particular, it seems reasonable to rank a person higher when that person has been nominated by another highly ranked person. Note that this is analogous to the PageRank metric discussed in Chapter 8. The idea as used in social networks is quite simple and brings us to the following definition of ranked prestige:

Definition 9.3: *Consider a simple directed graph D with vertex set $\{1, 2, \ldots, n\}$ with adjacency matrix $\mathbf{A}$ (i.e., $\mathbf{A}[i,j] = 1$ if and only if there is an arc $\overrightarrow{\langle i,j \rangle}$). The* **ranked prestige** *of a vertex k is defined as:*

$$p_{rank}(k) \stackrel{\text{def}}{=} \sum_{i=1, i \neq k}^{n} \mathbf{A}[i,k] \cdot p_{rank}(i)$$

9.2. SOME BASIC CONCEPTS

ID	Distance from j to i																$p_{prox}(v)$
	1	2	4	5	7	9	11	12	13	14	15	16	17	19	20	23	
1	0	4	3	1	4	5	3	1	2	2	3	2	4	2	1	4	0.366
2	4	0	4	5	2	1	2	3	5	6	1	3	3	1	2	2	0.341
4	2	5	0	3	5	6	4	1	1	4	4	1	5	3	2	5	0.294
5	1	5	2	0	5	6	4	2	1	1	4	2	5	3	2	5	0.313
7	5	1	5	6	0	2	1	4	6	7	2	4	2	2	3	1	0.294
9	5	1	5	6	1	0	1	4	6	7	2	4	2	2	3	2	0.294
11	5	1	5	6	2	2	0	4	6	7	1	4	1	2	3	2	0.294
12	1	4	2	2	4	5	3	0	3	3	3	1	4	2	1	4	0.357
13	2	5	1	3	5	6	4	1	0	4	4	1	5	3	2	5	0.294
14	1	4	3	1	4	5	3	2	2	0	3	2	4	2	1	4	0.366
15	4	2	4	5	1	3	2	3	5	6	0	3	2	1	2	1	0.341
16	2	4	1	3	4	5	3	1	2	4	3	0	4	2	1	4	0.349
17	4	2	4	5	3	3	1	3	5	6	2	3	0	1	2	1	0.333
19	3	2	3	4	2	3	1	2	4	5	1	2	2	0	1	2	0.405
20	2	3	2	3	3	4	2	1	3	4	2	1	3	1	0	3	0.405
23	4	2	4	5	3	3	1	3	5	6	2	3	1	1	2	0	0.333

Figure 9.8: Computing the proximity prestige for the classroom example. Each cell *(row, column)* denotes the distance from *column* to *row*.

Note that in order to compute $p_{rank}(k)$, we need to compute the ranked prestige of every vertex. Fortunately, the above equation is one of a total of n (one for each vertex), giving rise to a set of n equations in n unknowns. Standard mathematical techniques can be applied to solve these equations, although for even relatively small values of n, using software packages comes in handy. To illustrate the principle, let us consider a small social network with only three people A, B, and C. Each person is asked to give a weight $0 \leq w \leq 1$ to the other two, expressing the relative preference of one person over the other. So, for example, if A prefers B over C, she may express this by assigning a weight of 0.7 to B and 0.3 to C. Likewise, if B has no preference for either A or C, he should assign a weight of 0.5 to both of them. Note that the total weight that a person can assign to the others is always equal to 1. Let's assume that the weights have been assigned as follows:

ID	A	B	C
A	—	0.5	0.4
B	0.1	—	0.6
C	0.9	0.5	—

where we use the same notation as in Figure 9.8: cell (i, j) denotes the weight assigned *by* person j *to* person i. We now need to solve the following equations:

$$\begin{aligned} p_{rank}(A) &= 0.5 \cdot p_{rank}(B) + 0.4 \cdot p_{rank}(C) \\ p_{rank}(B) &= 0.1 \cdot p_{rank}(A) + 0.6 \cdot p_{rank}(C) \\ p_{rank}(C) &= 0.9 \cdot p_{rank}(A) + 0.5 \cdot p_{rank}(B) \end{aligned}$$

To simplify our notation a bit, we use the variables x, y, and z in place of $p_{rank}(A)$, $p_{rank}(B)$, and $p_{rank}(C)$, respectively. This then leads to:

$$\begin{aligned} x &= 0.5y + 0.4z \quad (1) \\ y &= 0.1x + 0.6z \quad (2) \\ z &= 0.9x + 0.5y \quad (3) \end{aligned}$$

If we would try to solve this set of equations, we would find only dependencies between x, y, and z. This is caused by the fact that we require that the sum of the values per column is always 1. In particular, by substituting (2) into (3), we find that $z = \frac{19}{14}x$. Likewise, by substituting (3) into (2), we find that $y = \frac{32}{35}x$. It is common practice to ensure that

$$\sqrt{\sum (p_{rank}(i))^2} = 1$$

which in our example would mean that

$$x^2 + \left(\frac{19}{14}x\right)^2 + \left(\frac{32}{35}x\right)^2 = 1$$

which, in turn, leads to:

$$x = 0.52 \qquad y = 0.48 \qquad z = 0.71$$

These values now express the ranked prestige of A, B, and C, respectively.

Note 9.2 (More information)
What we have actually been doing is computing what is known as an **eigenvector**. To explain, let $\mathbf{W}$ denote the matrix of nonnegative weights assigned between $n > 1$ people, such that $\mathbf{W}[i,j]$ is the weight assigned by person j to i. As in our example, we require that for each person j, $\sum_{i=1}^{N} \mathbf{W}[i,j] = 1$ and that $\mathbf{W}[j,j] = 0$. Let $\mathbf{p}$ be the vector of ranked prestiges:

$$\mathbf{p} \equiv (p_1, p_2, \ldots, p_n) \stackrel{\text{def}}{=} (p_{rank}(1), p_{rank}(2), \ldots, p_{rank}(n))$$

Using the abbreviation $w_{ij} = \mathbf{W}[i,j]$, we need to solve the set of equations

$$\begin{aligned} w_{11}p_1 + w_{12}p_2 + \cdots + w_{1n}p_n &= p_1 \\ w_{21}p_1 + w_{22}p_2 + \cdots + w_{1n}p_n &= p_2 \\ &\vdots \\ w_{n1}p_1 + w_{n2}p_2 + \cdots + w_{nn}p_n &= p_n \end{aligned}$$

9.2. SOME BASIC CONCEPTS

> which can be more concisely written in matrix form as
>
> $$\begin{pmatrix} w_{11} & w_{12} & \cdots & w_{1n} \\ w_{21} & w_{22} & \cdots & w_{2n} \\ \vdots & \vdots & & \vdots \\ w_{n1} & w_{n2} & \cdots & w_{nn} \end{pmatrix} \begin{pmatrix} p_1 \\ p_2 \\ \vdots \\ p_n \end{pmatrix} = \begin{pmatrix} p_1 \\ p_2 \\ \vdots \\ p_n \end{pmatrix}$$
>
> or, equivalently
>
> $$\mathbf{W} \cdot \mathbf{p} = \mathbf{p}$$
>
> In mathematical terms, **p** is the **eigenvector** that corresponds with the **eigenvalue** 1. As mentioned above, we generally require that $\sqrt{\sum(p_i)^2} = 1$, so that we can often find a unique solution for an eigenvector. For social network analysis, this eigenvector corresponds to the ranked prestiges.
>
> In general, eigenvectors are computed by first finding solutions to the more general equation
>
> $$\mathbf{W} \cdot \mathbf{p} = \lambda \mathbf{p}$$
>
> with λ being a scalar. Several solutions may exist, each known as an eigenvalue. In our case, because we demand that $\sum_i w_{ij} = 1$, one can show that the largest eigenvalue is $\lambda = 1$. We will not go into this material any further. A good introduction can be found in [Williams, 2001].

Let us finally see how we can compute the ranked prestige for each of the children in our classroom example. Again, we concentrate on the strongly connected component, consisting of 16 children. We need to construct a matrix that reflects the weight that child j assigns to child i. We follow two approaches. First, we consider the positive nominations and assign an equal weight to each nomination given by the same child. In other words, if A has nominated three other children, we assume that each of these three has the same influence on A. From Figure 9.8, we can seen that each child within the strongly connected component nominates exactly three other children in the same component, so that every weight is equal to $\frac{1}{3}$. In that case, the ranked prestige turns out to be as follows:

Child:	1	2	4	5	7	9	11	12
Ranked pres.:	0.148	0.171	0.132	0.056	0.123	0.057	0.332	0.369
Child:	13	14	15	16	17	19	20	23
Ranked pres.:	0.062	0.018	0.313	0.332	0.179	0.433	0.434	0.205

Our second approach entails the distance between children. In particular, reconsider the graph representing the positive nominations shown in Figure 9.7. We now take the distance from child i to child j (in this graph) as an indication of the how highly i ranks j. In particular, the larger the distance, the lower the ranking. Let M be the maximum eccentricity between two children in the largest strongly connected component. From our previous observations, we know that $M = 7$. If $d(i,j)$ denotes the shortest distance from child i to j, we define the weight w_{ij} that i assigns to j as:

$$w_{ij} \stackrel{\text{def}}{=} \frac{M - d(i,j)}{\sum\limits_{j \in R^-(i)} (M - d(i,j))}$$

Using these weights, we can then compute the ranked prestiges as:

Child:	1	2	4	5	7	9	11	12
Ranked pres.:	0.240	0.253	0.230	0.187	0.238	0.198	0.286	0.282
Child:	13	14	15	16	17	19	20	23
Ranked pres.:	0.195	0.134	0.282	0.279	0.245	0.315	0.311	0.252

Before we come to conclusions, we summarize our findings for the classroom in Figure 9.9. We also show the normalized values, obtained by dividing the measured importance by the found maximum importance for a specific metric. What we see is that different metrics lead to sometimes very different results. For example, the relative importance of children #4 and #5 depends on which metric we use: in the case of betweenness #5 is more important than #4, but this changes when ranked prestige as metric. Furthermore, it appears that ranked prestige generally leads to a greater variation (which is good). All metrics show the importance of children #19 and #20.

Structural balance

As stated by Wasserman and Faust [1994], a first important result from social network analysis was the theory of **structural balance**. The theory considers the sentiment relationships between people within a group, which are commonly modeled as *positive* of *negative*. In particular, the theory is concerned with examining whether the relationships between people are such that the group as a whole can be considered stable, or in balance. In its simplest form, the theory considers triads, that is, groups of three people. We briefly discussed triads and balance in Section 9.1 and will consider it in more detail here.

Let us first start with precisely defining balance. To this end, we need the definition of a signed graph:

Definition 9.4: *A **signed graph** is a simple graph G in which each edge is labeled with either a positive ("+") or negative ("−") sign. We denote the sign of an edge e as sign(e).*

9.2. SOME BASIC CONCEPTS

Child	Eccentricity	Closeness	Betweenness	Proximity prestige	Ranked prestige 1	Ranked prestige 2
1	5 (0.714)	0.023 (0.767)	0.140 (0.299)	0.366 (0.904)	0.148 (0.341)	0.240 (0.762)
2	6 (0.857)	0.021 (0.700)	0.153 (0.326)	0.341 (0.842)	0.171 (0.394)	0.253 (0.803)
4	6 (0.857)	0.018 (0.600)	0.050 (0.107)	0.294 (0.726)	0.132 (0.304)	0.230 (0.730)
5	4 (0.571)	0.025 (0.833)	0.105 (0.224)	0.313 (0.773)	0.056 (0.129)	0.187 (0.594)
7	7 (1.000)	0.018 (0.600)	0.083 (0.177)	0.294 (0.726)	0.123 (0.283)	0.238 (0.756)
9	7 (1.000)	0.018 (0.600)	0.007 (0.015)	0.294 (0.726)	0.057 (0.131)	0.198 (0.629)
11	7 (1.000)	0.018 (0.600)	0.155 (0.330)	0.294 (0.726)	0.332 (0.765)	0.286 (0.908)
12	5 (0.714)	0.022 (0.733)	0.220 (0.469)	0.357 (0.881)	0.369 (0.850)	0.282 (0.895)
13	6 (0.857)	0.018 (0.600)	0.016 (0.034)	0.294 (0.726)	0.062 (0.143)	0.195 (0.619)
14	3 (0.429)	0.030 (1.000)	0.054 (0.115)	0.366 (0.904)	0.018 (0.041)	0.134 (0.425)
15	6 (0.857)	0.021 (0.700)	0.083 (0.177)	0.341 (0.842)	0.313 (0.721)	0.282 (0.895)
16	5 (0.714)	0.021 (0.700)	0.140 (0.299)	0.349 (0.862)	0.332 (0.765)	0.279 (0.886)
17	6 (0.857)	0.021 (0.700)	0.017 (0.036)	0.333 (0.822)	0.179 (0.412)	0.245 (0.778)
19	5 (0.714)	0.025 (0.833)	0.466 (0.994)	0.405 (1.000)	0.433 (0.998)	0.315 (1.000)
20	4 (0.571)	0.025 (0.833)	0.469 (1.000)	0.405 (1.000)	0.434 (1.000)	0.311 (0.987)
23	6 (0.857)	0.021 (0.700)	0.029 (0.062)	0.333 (0.822)	0.205 (0.472)	0.252 (0.800)

Figure 9.9: Summary of the importance measures for the classroom example, with the normalized values shown between brackets.

A signed graph can be undirected or directed. For a signed graph G, we will use the notation $E^+(G)$ to denote the positive-signed edges and $E^-(G)$ for negative-signed edges.

The common interpretation of a positively signed edge between vertices A and B is that the two people represented by the vertices like each other. Analogously, a negative sign is to be interpreted as that they dislike each other. In the case of a signed directed graph, the likeness need not be symmetric. If A likes B, then this is represented by a positively signed arc $\overrightarrow{\langle A, B \rangle}$. A negatively signed arc from A to B means that A dislikes B. The absence of an arc (or edge in the case of an undirected graph) implies that two people neither like nor dislike each other. In the following, we will concentrate only on undirected signed graphs.

In Figure 9.4 we discussed how the various combinations of liking and disliking between people in a triad would lead to an (im)balanced situation. It can be readily seen that the balanced situation of a triad occurs if and only if there are zero or an even number of negative signed edges. This observation is generalized as follows:

Definition 9.5: *Consider an undirected signed graph G. The **product of two signs** s_1 and s_2 is again a sign, denoted as $s_1 \cdot s_2$. It is negative if and only if exactly one of s_1 and s_2 is negative. The **sign of a trail** T is the product of the signs of its edges: $sign(T) = \Pi_{e \in E(T)} sign(e)$.*

Note that the effect of multiplying signs can be easily understood if we substitute $+1$ for "$+$" and -1 for "$-$."

Note 9.3 (Mathematical language)
By now, you should be used to the fact that from time to time new mathematical symbols find their way into the text. In the previous definition, we have used the symbol "Π" as an abbreviation for multiplication, analogously to using the summation sign "$\sum$." In particular, we have

$$\Pi_{i=1}^{n} x_i \stackrel{\text{def}}{=} x_1 \times x_2 \times \cdots \times x_n$$

Note 9.4 (Mathematical language)
The definition of the product of a sign is a crude example of how mathematicians define what are known as (abstract) algebras. Algebras tell us how we can manipulate concepts such as signs, by providing basic rules concerning, for example, addition or multiplication. In the case of signs, we are interested only in multiplications. Adding more precision, we could have also included the following rules:

   Commutative: $s_1 \cdot s_2 = s_2 \cdot s_1$
   Associative: $(s_1 \cdot s_2) \cdot s_3 = s_1 \cdot (s_2 \cdot s_3)$

Note furthermore that the sign $I = $ "$+$" acts as an identity, i.e., for all signs s, we have that $I \cdot s = s \cdot I = s$. This same role of identity is played by the number "1" in our usual numbering systems.

A path (or cycle) is positive if it has zero or an even number of negative-signed edges. A negative-signed path (or cycle) is one that is not positive. We leave it as an exercise to prove the following theorem:

Theorem 9.1: *Consider an undirected signed graph G. For any trail T of G and $e \in E(T)$, $sign(T) = sign(e) \cdot sign(T - e)$.*

With these definitions at hand, we can now consider when sociograms that are represented as signed graphs are balanced:

9.2. SOME BASIC CONCEPTS

Definition 9.6: *An undirected signed graph is **balanced** when all its cycles are positive.*

An important characterization of a balanced graph is that its vertex set can be partitioned into two subsets such that all edges between the two subsets have negative sign, and no other edges. In other words, a group of people is balanced if it can be split into two subgroups such that members of the same subgroup like each other, yet members of different groups dislike each other (or don't care). This characterization was formally proven by Harary [1953], and is formalized by the following theorems.

Theorem 9.2: *An undirected signed complete graph G is balanced if and only if $V(G)$ can be partitioned into two disjoint subsets V_0 and V_1 such that each negative-signed edge is incident with a vertex from V_0 and one from V_1, and each positive-signed edge is incident with vertices from the same set. In other words:*

$$E^-(G) = \{\langle x,y \rangle | x \in V_0, y \in V_1\}$$
$$E^+(G) = \{\langle x,y \rangle | x,y \in V_0 \text{ or } x,y \in V_1\}$$

Proof. Assume that G is balanced. Let $u \in V(G)$ and let $N^+(u)$ consist of all vertices adjacent to u through a positive-signed edge. Set

$$V_0 \leftarrow \{u\} \cup N^+(u) \text{ and } V_1 \leftarrow V(G) \setminus V_0.$$

Consider two vertices $v_0, w_0 \in V_0$, other than u. Because the edges $\langle u, v_0 \rangle$ and $\langle u, w_0 \rangle$ have positive signs, and because G is balanced, we must also have that $\langle v_0, w_0 \rangle$ has a positive sign (note that edge $\langle v_0, w_0 \rangle$ exists because G is a complete graph). Likewise, consider any two vertices $v_1, w_1 \in V_1$. Again, because G is balanced, we know that the triad with vertices u, v_1, w_1 must be positive, and because edges $\langle u, v_1 \rangle$ and $\langle u, w_1 \rangle$ have negative signs, edge $\langle v_1, w_1 \rangle$ must have a positive sign. Finally, consider the edge $\langle v_0, v_0 \rangle$, which is part of the triad with vertices u, v_0 and v_1. With the sign of $\langle u, v_0 \rangle$ being positive and that of $\langle u, v_1 \rangle$ negative, and G being balanced, edge $\langle v_0, v_1 \rangle$ must have a negative sign. We conclude that V_0 and V_1 partition $V(G)$ as required.

Conversely, assume that $E^-(G)$ and $E^+(G)$ satisfy the stated conditions. Every cycle in G contains an even number of edges from $E^-(G)$, implying that the sign of every cycle is positive. By definition, G is balanced. □

Note 9.5 (Study tip)
The proof of Theorem 9.2 is much easier to understand when using a drawing. As mentioned before, studying graph theory generally requires you to visualize situations by sketching graphs. Do the same for this proof.

We leave it as an exercise to show that every subgraph of a balanced signed graph is again balanced. We will need this property for the following theorem:

Theorem 9.3: *Consider an undirected signed graph G and two distinct vertices $u, v \in V(G)$. G is balanced if and only if all (u, v)-paths have the same sign.*

Proof. First assume that G is balanced. Let P and Q be two distinct (u, v)-paths. Consider the set of edges E' obtained from P and Q after removing the ones they have in common, that is

$$E' = (E(P) \cup E(Q)) \setminus (E(P) \cap E(Q)).$$

What can we say about the subgraph H induced by E'? First note that there can be no cycles having edges in common. If that were the case, those common edges would have been part of both P and Q, which by construction cannot happen. In other words, any two cycles in H have no edges in common. Because H is a subgraph of G, it must also be balanced. As a consequence, all cycles in H are positive. Furthermore, each cycle C in H consists of exactly two subpaths $\hat{P}$ from P and $\hat{Q}$ from Q. That is, $E(C) = E(\hat{P}) \cup E(\hat{Q})$. Because $\hat{P}$ and $\hat{Q}$ have no edges in common, and because $sign(C) = sign(\hat{P}) \cdot sign(\hat{Q})$ is positive, we conclude that the signs of $\hat{P}$ and $\hat{Q}$ must be the same. Taking all cycles of H into account, along with the edges common to both P and Q, we conclude that P and Q must have the same sign.

Conversely, assume that (u, v)-paths have the same sign. Because u and v have been chosen arbitrarily, and because every cycle C can be constructed as the union of two edge-disjoint paths P and Q, we necessarily have that $sign(C) = sign(P) \cdot sign(Q)$ must be positive. Hence, G is balanced. □

Combining theorems now allows us to prove the following general characterization of balanced signed graphs, again due to Harary [1953].

Theorem 9.4: *An undirected signed graph G is balanced if and only if $V(G)$ can be partitioned into two disjoint subsets V_0 and V_1 such that the following two conditions hold:*

(1) $E^-(G) = \{\langle x, y \rangle | x \in V_0, y \in V_1\}$
(2) $E^+(G) = \{\langle x, y \rangle | x, y \in V_0 \text{ or } x, y \in V_1\}.$

Proof. First, let us assume that G is balanced. Without loss of generality, we also assume that G is connected. The theorem is proven to hold by induction on the number m of edges of G. Clearly, the theorem is seen to hold

9.2. SOME BASIC CONCEPTS

for the case that $m = 1$, so assume it holds for $m > 1$. Consider any two nonadjacent vertices u and v of G. From the previous theorem, we know that all (u,v)-paths have the same sign. Therefore, extend G by adding the edge $e = \langle u, v \rangle$ with the same sign as any (u,v)-path in G, leading to the new graph $G^* = G + e$. Any newly introduced cycle C in G^* will consist of e and a (u,v)-path P from G. Because $sign(C) = sign(e) \cdot sign(P)$, and $sign(e) = sign(P)$, C must be positive, and thus the extended graph is also balanced. Continue in this way with adding edges between nonadjacent vertices until we have a signed *complete* graph G^{**}, which we know is balanced. From Theorem 9.2, it follows that we can partition the vertex set of G^{**}, and thus also G into the two required subsets.

Conversely, assume we can partition G into two subsets V_0 and V_1 as described. Extend G by adding an edge $e = \langle u, v \rangle$ between two nonadjacent vertices, leading to $G^* = G + e$. If u and v lie in the same subset, $sign(e)$ becomes positive, otherwise negative. Continue in this way adding edges until we have a signed complete graph G^{**}. Again, from Theorem 9.2 we know that this graph is balanced, and because G is a subgraph of G^{**}, we know G is also balanced. □

With this characterization, it is now relatively easy to check whether a signed graph is balanced. The following algorithm will do the trick.

Algorithm 9.1 (Balanced graphs): *Consider an undirected, connected signed graph G. For any vertex $v \in V(G)$, denote by $N^+(v)$ the set of vertices adjacent to v through a positive-signed edge, and by $N^-(v)$ the set of vertices adjacent through a negative-signed edge. Let I be the set of inspected vertices so far.*

1. *Select an arbitrary vertex $u \in V(G)$ and set $V_0 \leftarrow \{u\}$ and $V_1 \leftarrow \emptyset$. Set $I \leftarrow \emptyset$.*

2. *Select an arbitrary vertex $v \in (V_0 \cup V_1) \setminus I$. Assume $v \in V_i$.*
   - *For all $w \in N^+(v) : V_i \leftarrow V_i \cup \{w\}$.*
   - *For all $w \in N^-(v) : V_{(i+1) \bmod 2} \leftarrow V_{(i+1) \bmod 2} \cup \{w\}$.*
   - *Also, $I \leftarrow I \cup \{v\}$.*

3. *If $V_0 \cap V_1 \neq \emptyset$ stop: G is not balanced. Otherwise, if $I = V(G)$ stop: G is balanced. Otherwise, repeat the previous step.*

Note 9.6 (Mathematical language)
For the previous algorithm we have used a concise notation that may require some effort to understand:

$$V_{(i+1) \bmod 2} \leftarrow V_{(i+1) \bmod 2} \cup \{w\}.$$

> Note that we have assumed that the arbitrarily selected vertex v is in set V_i. As a consequence, when $v \in V_0$, $V_{(i+1) \bmod 2}$ is equal to V_1, whereas for $v \in V_1$, we see that $V_{(i+1) \bmod 2}$ is equal to $V_{2 \bmod 2} = V_0$. In other words, $V_{(i+1) \bmod 2}$ refers to the *other* set than the one containing v.

To see why this algorithm is correct, first note that if may be possible for a vertex to be added to V_0 and later also to V_1 (or *vice versa*). Whenever this happens, we will not be able to partition the vertex set anymore as is required for a signed graph to be balanced. In step 3, we will decide to stop inspecting (uninspected) vertices from $V_0 \cup V_1$ if the two sets are not disjoint anymore, or until each vertex has been placed in either V_0 or V_1, at which point it must be the case that $V_0 \cap V_1 = \emptyset$, so that G is indeed balanced.

Cohesive subgroups

Given a social network, researchers have always been keen on identifying groups of closely bound people, or better known as **cohesive subgroups**. Typical examples of such groups in practice are formed by families and friends. More recent, interest has grown in identifying groups of, for example, terrorists. And although it seems naturally evident what a cohesive subgroup actually entails, formalizing the concept in graph theory such that it matches what one expects in real life is less obvious. Let us take a look at a few proposals (see also [Mokken, 1979]).

One of the earliest proposals for modeling cohesive subgroups was to consider (maximal) cliques:

Definition 9.7: *Consider an undirected simple graph G. A **(maximal) clique** of G is a complete subgraph H of at least three vertices such that H is not contained in a larger complete subgraph of G. A clique with k vertices is called a **k-clique**.*

Note that a graph can have several cliques. Consider, for example, the graph in Figure 9.10. In this case, we see that there are two cliques: the 3-clique induced by the set of vertices $\{2, 4, 5\}$ and the 4-clique induced by $\{1, 2, 3, 5\}$. This example also shows that a vertex may be contained in two different cliques.

The problem with using cliques as a means for modeling cohesive subgroups is that they are generally too restrictive. In the first place, many subgroups exist in reality in which not all members relate to each other. In terms of graphs, this means that that a subgroup cannot always be adequately represented by a complete subgraph. Related to this strictness is that by considering only cliques, it turns out that only small subgroups can

9.2. SOME BASIC CONCEPTS

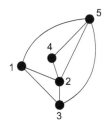

Figure 9.10: A graph with two maximal cliques.

be identified. Considering that in many cases sociograms are based on questionnaires in which people are asked to identify their k best relations, we also see that the degree of a vertex can never be more than k, and thus that a maximal clique can have only $k+1$ members. With such restrictions, it may even be impossible to identify any clique.

For these reasons, researchers have been looking for other metrics for defining subgroups. One approach is to relax how strong the bonds between members of a subgroup should be. In particular, one can also define a subgroup as the maximal subgraph in which the distance between its members is less or equal to a constant k. This leads to what are known as k-distance-cliques:

Definition 9.8: *Let G be an undirected simple graph. A **k-distance-clique** of G is a maximal subgraph H of G such that for all vertices $u, v \in V(H)$, the distance $d_G(u,v) \leq k$.*

(We have introduced this term to avoid confusion with k-cliques. Note, however, that k-distance-cliques are often also referred to as k-cliques [Scott, 2000; Wasserman and Faust, 1994].) It is important to note that the distance between two vertices in a k-distance-clique is measured relative to the original graph G, as is indicated by the notation $d_G(u,v)$. This means that two vertices u and v in a k-distance-clique H may be connected through a shortest path in H that is *longer* than a shortest (u,v)-path in G. This implies that the diameter of a k-distance-clique may be larger than k, which is somewhat counter-intuitive. Another problem with k-distance-cliques is also caused by the fact that distance is measured with respect to the original graph: it is possible to construct a graph in which a k-distance-clique may be disconnected (see exercises). To ensure that the diameter of a subgraph matches one's intuition, Mokken [1979] proposed k-clans:

Definition 9.9: *Let G be an undirected simple graph. A **k-clan** of G is a k-distance-clique H of G such that for all vertices $u, v \in V(H)$, the distance $d_H(u,v) \leq k$.*

The only, yet important, difference with k-distance-cliques is that distance is measured relative to H instead of G. By definition, every k-clan is also a k-distance-clique. If we take the diameter as the sole criterion, we obtain what are known as k-clubs:

Definition 9.10: *Let G be an undirected simple graph. A **k-club** of G is a maximal subgraph H of G such that diam(H) $\leq k$. In other words, max$\{d_H(u,v)|u,v \in V(H)\} \leq k$.*

We will show that every k-clan of a graph G is also a k-club of G. However, not every k-club is also a k-clan, as can be seen from Figure 9.11. In this example, we have two 2-distance-cliques: $H_1 = G[\{1,2,3,5,6\}]$ and $H_2 = G[\{2,3,4,5,6\}]$. H_2 is also a 2-club, as well as a 2-clan. In addition, both $H_3 = G[\{1,2,5,6\}]$ and $H_4 = G[\{1,2,3,6\}]$ are 2-clubs, but neither are 2-distance-cliques, and thus are not 2-clans.

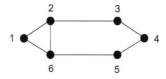

Figure 9.11: Graph illustrating cliques, clans, and clubs.

Now consider a k-club H of a graph G. Because for all vertices $u,v \in V(H)$, we know that $d_G(u,v) \leq d_H(u,v) \leq k$, H must be contained in a k-distance-clique of G. We use this property to prove the following:

Theorem 9.5: *Every k-clan of a graph G is also a k-club.*

Proof. From the definitions of k-clan and k-club, one can easily see that for a k-clan H we certainly have that for all vertices $u,v \in V(H)$, $d_H(u,v) \leq k$. Therefore, we merely need to show that H is also maximal with respect to the definition of a k-club. To this end, assume that H is not maximal. This means that there is a set of vertices $S \subset V(G)\setminus V(H)$ such that for all $u \in V(H)$ and $s,t \in S$, we have:

$$d_G(u,t) \leq d_{H^*}(u,t) \leq k \quad \text{and} \quad d_G(s,t) \leq d_{H^*}(s,t) \leq k$$

where $H^* = G[V(H) \cup S]$. However, because H is also a k-distance-clique, this would violate the maximality of H as a k-distance-clique, contradicting our assumption of the existence of S. Hence, H is also maximal as a k-club, completing the proof. □

9.2. SOME BASIC CONCEPTS

The real problem with these definitions is that all of them are still very strict when it comes to selecting whether a vertex belongs to a group or not. In reality, cohesiveness of a group is much more fuzzy: if Alice considers Bob to be her best friend, it may very well be the case that Bob's best friend Chuck is considered by Alice to be just an acquaintance of her. In other words, we would normally present a link between Alice and Chuck, but the meaning is different than the one between Alice and Bob. Such relationships can be captured through weighted graphs, but the definitions of cohesive groups do not cater for such situations.

In the same light, we could consider an alternative formulation of k-cliques by defining a group based on the minimal degree of each vertex:

Definition 9.11: *Let G be an undirected simple graph. A **k-core** of G is a maximal subgraph H of G such that for all vertices $u \in V(H)$, the degree $\delta(u) \geq k$.*

In other words, each vertex in a k-core is joined with at least k other member of that group. Again, it turns out that such a definition is often just too strict: it draws boundaries around groups that cannot account for the natural "exceptions to the rule."

A much better approach is to follow data-clustering techniques for identifying communities. As reported by Porter et al. [2009], a large variety of older and newer techniques have been proposed leading to much better results. Let us discuss one such method, known as **clique percolation** [Palla et al., 2005].

Clique percolation is based on identifying groups based on maximal cliques, yet with the important difference that groups may overlap. In other words, vertices may belong to different cliques without the necessity of having a maximal degree (as defined by the size of the clique it is member of). We can then define a k-clique community:

Definition 9.12: *Let G be an undirected simple graph. Two k-cliques C_1 and C_2 are said to be **adjacent** if they have at least $k-1$ vertices in common: $|V(C_1) \cap V(C_2)| = k-1$. A **k-clique community** of G is a union of k-cliques $\mathbf{C} = \{C_1, \ldots, C_n\}$ such that for every two k-cliques $C_u, C_v \in \mathbf{C}$, there is a series $[C_u = C_{u_0}, C_{u_1}, \ldots, C_{u_m} = C_v]$ in which C_{u_i} and $C_{u_{i+1}}$ are adjacent k-cliques of $\mathbf{C}$.*

This definition is best understood by taking a look at an example. Let's consider our social network of Figure 9.1, which we show again in Figure 9.12 along with the various 3-cliques and single 4-clique. Note that in this example there are no k-cliques for $k \geq 5$. What can we say about the adjacency of cliques? First, it is not difficult to see that our single 4-clique, denoted C_1, is not adjacent to any other clique for the simple reason that it does not have a single vertex in common with any one of them. Likewise, if we consider 3-cliques C_7 and C_8, we see that Sam is member of both of them.

However, because Sam is the *only* member that is shared between the two cliques, they are not considered to be adjacent: two 3-cliques are adjacent only if they share two vertices. For the same reason, we see that 3-cliques C_8, C_9, and C_2 are not adjacent to any other clique.

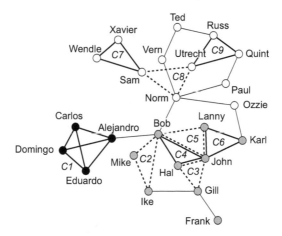

Figure 9.12: The social network from Figure 9.1, showing the various k-cliques.

The story is different for cliques C_3 and C_4: because $V(C_3) \cup V(C_4) = \{\text{Hal, John}\}$, the two are adjacent. In fact, C_3, C_4, C_5, and C_6 form a 3-clique community, as shown in Figure 9.13. We see that besides C_3 and C_4 that also C_4 and C_5, as well as C_5 and C_6 are pairs of adjacent 3-cliques. The result is that using this method of identifying cohesive groups, we find ourselves dealing with six communities: $\{C_1\}$, $\{C_2\}$, $\{C_3, C_4, C_5, C_6\}$, $\{C_7\}$, $\{C_8\}$, and $\{C_9\}$.

	C_3	C_4	C_5	C_6
C_3	—	{Hal, John}	$C_3-C_4-C_5$	$C_3-C_4-C_5-C_6$
C_4	{Hal, John}	—	{Bob, John}	$C_4-C_5-C_6$
C_5	$C_3-C_4-C_5$	{Bob, John}	—	{Lanny, John}
C_6	$C_3-C_4-C_5-C_6$	$C_4-C_5-C_6$	{Lanny, John}	—

Figure 9.13: A 3-clique community. Every entry shows either the intersection between two adjacent 3-cliques, or the path of 3-cliques between two nonadjacent cliques.

9.2. SOME BASIC CONCEPTS

Note 9.7 (More information)
Palla et al. [2007] have extended clique percolation to directed graphs. In the undirected case, a clique represents a maximal group in which all vertices are considered equally important. In a directed graph, we need to account for the fact that relations are no longer symmetric, but that they reflect some ordering between vertices. For this reason Palla et al. have been looking for an ordering of the vertices in their definition of a directed k-clique. In the following, we use the notation $u \prec v$ to indicate that vertex u precedes vertex v in an ordering of vertices.

Definition 9.13: *Consider a directed graph D. A **directed k-clique** is a directed subgraph H with k vertices such that (1) the underlying graph of H is complete, and (2) there is an ordering of the vertices of H, such that if $u \prec v$ then $\langle \overrightarrow{u,v} \rangle \in A(H)$.*

To illustrate, consider directed acyclic graphs, which we encountered in Chapter 3. In this case, for a directed clique H, a natural ordering of vertices can be found by considering the outdegree of each vertex. In particular, $u \prec v$ if u's outdegree (in H) is larger than v's. It can be shown that in this case such an ordering always exists. To illustrate, Figure 9.14 shows how we can come to such an ordering a directed acyclic graph.

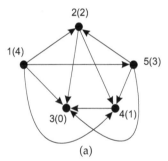

Position	Vertex	$\langle \overrightarrow{u,v} \rangle \in A$?
1	1	$\langle \overrightarrow{1,5} \rangle \in A$
2	5	$\langle \overrightarrow{5,2} \rangle \in A$
3	2	$\langle \overrightarrow{2,4} \rangle \in A$
4	4	$\langle \overrightarrow{4,3} \rangle \in A$
5	3	irrelevant

(a)   (b)

Figure 9.14: (a) A (complete) directed acyclic graph. The outdegree of each vertex is shown as well. An ordering of the vertices is shown in (b).

To examine directed subgraphs in which two vertices u and v are mutually joined (i.e., both $\langle \overrightarrow{u,v} \rangle, \langle \overrightarrow{v,u} \rangle \in A(H)$), we merely need to remove one the arcs from either u to v or from v to u. In many cases, the remaining subgraph will be acyclic, in which case we can use the ordering based on a vertex's outdegree. There may also be cases in which an ordering cannot be found, meaning that we are not dealing with a directed k-clique.

Again, two directed k-cliques are considered adjacent if they share $k - 1$ vertices. Then, using these definitions, it turns out that for our classroom example shown in Figure 9.7(a), the directed critical percolation method will find exactly two directed 3-communities: one consisting of all the girls, and one consisting of all the boys. None of the methods we have discussed so far would have been

> capable of coming to such an identification of subgroups. Further information on critical percolation for directed graphs can be found in [Palla et al., 2007].

Affiliation networks

As a last example of important concepts in social networks, we consider what are known as **affiliation networks** [Wasserman and Faust, 1994; Knoke and Yang, 2008]. In such a network, people are tied to each other through a membership relation. For example, Alice and Bob may be member of the same sportsclub, or are both member of the same management team. In general, affiliation networks are constructed from a set of actors and a set of social events, where each actor is said to participate in one or several events. An affiliation network can be naturally represented as a bipartite graph, with each vertex representing either an actor or an event. An edge represents the participation of an actor in a specific event.

Affiliation networks have been studied for a variety of reasons, but two are particularly important for our discussion [Wasserman and Faust, 1994]. First, it is argued that there is a lot of information to discover between individuals by considering the events that they share, and likewise, correlation between events can be discovered by considering the shared participation by actors. In other words, the indirect relationship between individuals that is caused by the events they share is an important object of study, and the same holds for the indirect relationship between two events caused by individuals participating in both events.

The second reason is that sociologists believe that participation in common events helps to explain the existence of ties between two individuals. For example, it is believed that influence patterns are established by the fact that people participate in shared events. As a consequence, understanding how information is diffused, or how innovations are adopted, may require an understanding of shared events between people.

Because affiliation networks consist of two different sets, they are also referred to as **two-mode networks**. However, when considering the two main reasons for studying them, we see that they are effectively used to study the (indirect) relationships between individuals or events. This brings us back to our original conception of social networks, now referred to as **one-mode networks**.

Let us first consider the adjacency matrix representing an affiliation network. Let V_A denote the set of vertices representing the actors, and V_E the set representing events. We consider only the (actor, event) submatrix **AE** consisting of $n_A = |V_A|$ rows and $n_E = |V_E|$ columns. Clearly, we

9.2. SOME BASIC CONCEPTS

have that $\mathbf{AE}[i,j] = 1$ if and only if actor i participates in event j. Furthermore, $\sum_{i \in V_A} \mathbf{AE}[i,j]$ tells us how many actors participate in event j, whereas $\sum_{j \in V_E} \mathbf{AE}[i,j]$ tells us in how many events actor i participates.

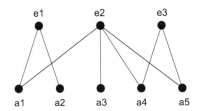

	e1	e2	e3
a1	1	1	0
a2	1	0	0
a3	0	1	0
a4	0	1	1
a5	0	1	1

Figure 9.15: An example affiliation network with adjacency submatrix.

Let us consider the simple affiliation network shown in Figure 9.15, along with its adjacency submatrix. Now consider the following sum:

$$\mathbf{NE}[i,j] = \sum_{k=1}^{n_E} \mathbf{AE}[i,k] \cdot \mathbf{AE}[j,k]$$

Note that $\mathbf{AE}[i,k] \cdot \mathbf{AE}[j,k] = 1$ if and only if both actors i and j participated in event k. In other words, $\mathbf{NE}[i,j]$ counts the number of events in which both actor i and j participated. Likewise, we can compute:

$$\mathbf{NA}[i,j] = \sum_{k=1}^{n_A} \mathbf{AE}[k,i] \cdot \mathbf{AE}[k,j]$$

in which case we are counting the number of actors participating in both event i and j. Note that $\mathbf{AE}[k,i] \cdot \mathbf{AE}[k,j] = 1$ if and only if actor k participated in both events i and j. The values for these two tables are shown in Figure 9.16. Of course, for both tables we have:

$$\mathbf{NE}[i,j] = \mathbf{NE}[j,i] \quad \text{and} \quad \mathbf{NA}[i,j] = \mathbf{NA}[j,i]$$

Furthermore, it is not difficult to see that $\mathbf{NE}[i,i] = \delta(a_i)$ and $\mathbf{NA}[i,i] = \delta(e_i)$.

How does this work in practice? In 2006 a major Dutch newspaper conducted an investigation to identify the most influential people within the Netherlands [Dekker and van Raaij, 2006]. The research was inspired by a statement in 1968 by Jan Mertens, at the time a union leader, that the Netherlands was effectively governed by approximately 200 people. Since 2006, identifying the top-200 most influential people has become a yearly returning event, with the not perhaps so surprising result that the top hasn't

NE	a1	a2	a3	a4	a5
a1	2	1	1	1	1
a2	1	1	0	0	0
a3	1	0	1	1	1
a4	1	0	1	2	2
a5	1	0	1	2	2

NA	e1	e2	e3
e1	2	1	0
e2	1	4	2
e3	0	2	2

Figure 9.16: The matrices **NE** and **NA** from Figure 9.15.

changed a lot. The core of the work is centered around a two-mode network, for which the technical setup and analysis is described in de Nooy [2006]. Actually determining which people are the most influential cannot be done by interpretation of raw network data. Instead, several metrics that have been described so far have been adjusted to more realistically reflect relationships. For example, rather than taking the distance as the length d of a shortest path, it was taken proportional to 2^d.

For our purposes, we take a simple approach and merely consider the largest connected component of the two-mode network of approximately 200 people. This leads to an affiliation network representing 197 actors and 391 events. An event is typically a board of directors, a supervisory board, etc. The graph is shown in Figure 9.17 where people are represented by boxes and events by circles.

Of course, by merely looking at this graph it is already very difficult to draw any conclusions. However, when we consider the matrices **NE** and **NA**, we see that more than 1250 pairs of actors share at least one event that both participate in. In particular, there is not a single actor who does *not* participate in at least one event with another actor. In fact, there are a number of actors who participate in at least three same events. When we take a look at the matrix **NA** we see that there is hardly any event for which its participants do not participate in another event. Apparently, it is common for the top to participate in at least two events. There is even a pair of events with as much as nine actors in common. One could argue that in such cases, participating in one event implicitly means that you'll be participating in the other as well.

9.3 Equivalence

So far, we have essentially been concentrating on identifying the properties of a specific person, or a group of persons, in a social network. An important, yet sometimes difficult question is identifying the *position* or *role* that

9.3. EQUIVALENCE

someone has. For social networks, answering such a question is related to identifying similarity between (groups of) people based on the structure of the network or structure of subnetworks. In this section, we will take a closer look at three related concepts that have been used for this purpose.

Structural equivalence

Consider the situation that in a social network two people, or actors A and B, have exactly the same relationships to the other actors in the network. In other words, if A is linked to C, then so is B, and if there is no link between A and D, then there is also no link between B and D. From the perspective of the network, you can argue that A and B are essentially indistinguishable: they apparently play the same role. This notion of similarity has called structural equivalence, first formally defined by Lorrain and White [1971]:

Definition 9.14: *Let D be a directed graph. Two vertices u and v are **structurally equivalent** if their respective sets of in-neighbors and out-neighbors are the same: $N_{in}(u) = N_{in}(v)$ and $N_{out}(u) = N_{out}(v)$.*

In other words, two vertices u and v are structurally equivalent if u has arcs to exactly the same vertices as v, but also all vertices that are linked

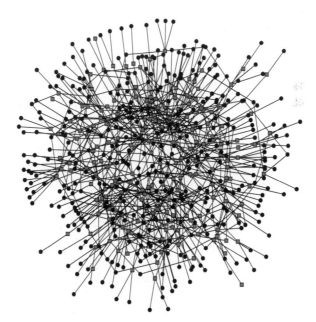

Figure 9.17: The graph of 2006 top-200 most influential people in The Netherlands.

to u are linked to v. Indeed, from the perspective of a network, vertices u and v are indistinguishable. Structural equivalence can easily be defined for undirected graphs as well, in which case we require that $N(u) = N(v)$. Figure 9.18 shows a simple social network with two structurally equivalent vertices u and v.

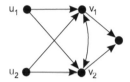

Figure 9.18: A simple social network with structural equivalent vertices u_1 and u_2.

The formal definition of structural equivalence is rather strict. For example, if u and v are each other's neighbor, then by definition they can never be structurally equivalent. For this specific situation, equivalence between two vertices u and v may exclude these two vertices from the respective sets of neighbors. In that case, vertices v_1 and v_2 from Figure 9.18 would also be structurally equivalent. But even then it is highly unlikely to see any two actors in practical situations to have exactly the same neighbors. For this reason it makes sense to not look for strict equivalence but to seek for a weaker form in which two vertices are "almost" equivalent. To this end we can define the following distance metric to express the extent that two vertices are the same.

Definition 9.15: *Consider a (strict) directed graph D with vertex set $V(D) = \{v_1, \ldots, v_n\}$ and adjacency matrix $\mathbf{A}$. The **Euclidean distance** $d(v_i, v_j)$ between two vertices v_i and v_j is defined as:*

$$d(v_i, v_j) \stackrel{\text{def}}{=} \sqrt{\sum_{k=1}^{n} \left((\mathbf{A}[i,k] - \mathbf{A}[j,k])^2 + (\mathbf{A}[k,i] - \mathbf{A}[k,j])^2 \right)}$$

Recall that for a strict directed graph, $\mathbf{A}[i,j] = 1$ if and only if there is an arc from v_i to v_j. As a consequence, $d(v_i, v_j) = 0$ if and only if vertices v_i and v_j are structurally equivalent: for each k, $\mathbf{A}[i,k] = \mathbf{A}[j,k]$ and $\mathbf{A}[k,i] = \mathbf{A}[k,j]$.

The Euclidean distance between two vertices now gives us a measure to see to what extent two vertices are structurally equivalent. Consider the graph shown in Figure 9.19(a). It is not difficult to see that v_1 and v_2 are structurally equivalent, but it would also appear that v_3 and v_4 are structurally very similar. If we compute the Euclidean distances, shown in Figure 9.19(b), we see that indeed v_3 and v_4 are relatively close to each other

9.3. EQUIVALENCE

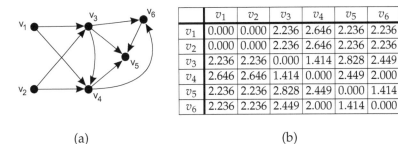

(a) (b)

Figure 9.19: (a) A directed graph and (b) the Euclidean distances between its vertices.

in comparison to other pairs of nonequivalent vertices. We leave it as an exercise to the reader to actually compute the various Euclidean distances.

Note 9.8
To get an impression of what the chances are of being structurally equivalent, let's consider a directed $ER(n,p)$ random graph for which p indicates the probability that there is an arc $\langle \vec{u},\vec{v} \rangle$ for an arbitrarily chosen pair of vertices u and v. The probability that two vertices u and v have an arc to the same vertex w, is obviously p^2. If both have outdegree k_{out}, then the probability that they have exactly the same set of out-neighbors is equal to $\binom{n-2}{k_{out}}(p^2)^{k_{out}}(1-p^2)^{n-2-k_{out}}$. Likewise, if they both have indegree k_{in}, the probability of having exactly the same set of in-neighbors is equal to $\binom{n-2}{k_{in}}(p^2)^{k_{in}}(1-p^2)^{n-2-k_{in}}$. Given the fact that even having the same vertex degree can be rather low, it is not hard to see that finding two structurally equivalent vertices in a directed graph is indeed very low. Therefore, the implication of finding such nodes in real networks means that something interesting may be going on.

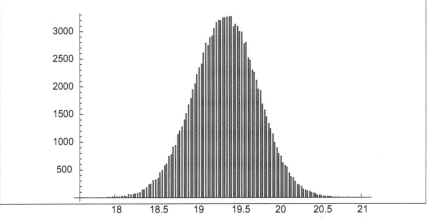

Figure 9.20: The distribution of distances in a directed $ER(500, 0.25)$ random graph.

	v_1	v_2	v_3	v_4	v_5	v_6
v_1	0.000	0.000	2.236	2.646	2.236	2.236
v_2	0.000	0.000	2.236	2.646	2.236	2.236
v_3	2.236	2.236	0.000	1.414	2.828	2.449
v_4	2.646	2.646	1.414	0.000	2.449	2.000
v_5	2.236	2.236	2.828	2.449	0.000	1.414
v_6	2.236	2.236	2.449	2.000	1.414	0.000

> As a further illustration, Figure 9.20 shows the distribution of Euclidean distances between pairs of vertices in a directed $ER(500, 0.25)$ random graph. We conclude that only very few vertices lie close to each other when taking the Euclidean distance as metric. Again, this means that if we do find vertices close to each other, then this should be treated as quite exceptional, which is exactly what we hope to find when looking for what could be called structural similarity.

Automorphic equivalence

As mentioned, structural equivalence is rather strict as it demands that the neighbor sets of two vertices are exactly the same. In effect, two structurally equivalent vertices are considered to be interchangeable and have the same *position* in a network. However, we are often looking for nodes in a social network that have similar *roles* (see also Wasserman and Faust [1994]). For example, we may want to identify who are teachers in a school. The basic assumption underlying such an identification is that we should look at the structure of the subgraph surrounding specific vertices. Indeed, this brings us to considering graph isomorphisms again, which we discussed in Section 2.2.

In particular, we are looking for a way to exchange two vertices, along with their respective neighbors, such that the resulting graph remains "the same." To make this more precise, recall first the definition of graph isomorphism:

Definition 9.16: *Consider two graphs $G = (V, E)$ and $G^* = (V^*, E^*)$. G and G^* are **isomorphic** if there exists a one-to-one mapping $\phi : V \to V^*$ such that for every edge $e \in E$ with $e = \langle u, v \rangle$, there is a unique edge $e^* \in E^*$ with $e^* = \langle \phi(u), \phi(v) \rangle$.*

Keeping a graph "the same" is essentially asking whether a graph is isomorphic with itself, but using a nontrivial remapping of vertices. Nontrivial means that at least some vertices are not mapped onto themselves. Formally, we speak of an automorphism, which is defined as follows:

Definition 9.17: *Consider an undirected graph $G = (V, E)$. An **automorphism** is a one-to-one mapping $\phi : V \to V$ such that for every edge $e \in E$ with $e = \langle u, v \rangle$, there is a unique edge $e^* \in E$ with $e^* = \langle \phi(u), \phi(v) \rangle$. An automorphism ϕ is called nontrivial if at least for one vertex $v \in V$ we have that $\phi(v) \neq v$.*

Note that the definition of automorphism can be easily extended to directed graphs. We can now define when two nodes in a social network play the same role by considering the associated (directed or undirected) graph:

9.3. EQUIVALENCE

Definition 9.18: *Consider a graph G. Two distinct vertices u and v are **automorphically equivalent** if and only if there is an automorphism ϕ for G with $\phi(u) = v$.*

To illustrate the idea of automorphical equivalence, consider the social network shown in Figure 9.21. In this example, it is not difficult to see that the two subgraphs H_1 and H_2 are not only isomorphic, but that they can also be "swapped" to obtain essentially the same graph. In particular, the mapping $\phi(u_i) = v_i$ will do the job. This also means that each pair of vertices (u_i, v_i) are automorphically equivalent. Finally, note that just as in the case of graph isomorphism, finding a (non trivial) automorphism may be a difficult task to accomplish.

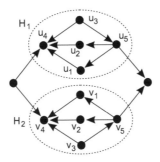

Figure 9.21: An example of a directed graph with automorphically equivalent vertices.

Regular equivalence

Both structural and automorphical equivalence have relatively simple graph-theoretical formulations, yet may be rather difficult to use in practice. As it turns out, for sociological research, another type of equivalence is often more important as it more naturally reflects the notion of a *role* [Hanneman and Riddle, 2005]: regular equivalence. Informally, two nodes in a social network are regularly equivalent if they fulfill the same role. The latter is decided by taking a look at the nodes to which the two nodes are linked: if the respective destinations are also regularly equivalent, then so are the sources. For example, two people may be identified as regularly equivalent because both have a link to two nurses, which had already been identified as being regularly equivalent. In this case, the two sources may turn out to be doctors.

An issue with this definition is that it is recursive: being regularly equivalent depends on the equivalence of the targets. Formally, we have:

Definition 9.19: *Let G be an undirected graph. Two vertices u_1 and u_2 are said to be **regularly equivalent** if for all edges $\langle u_1, v_1 \rangle \in E(G)$ there is an edge $\langle u_2, v_2 \rangle \in E(G)$ such that v_1 and v_2 are also regularly equivalent.*

Another way of looking at regular equivalence is coloring the vertices of a graph such that if two vertices u and v have the same color, then for each neighbor of u there will be a neighbor of v with the same color. Consider the graph shown in Figure 9.22(a), taken from Borgatti and Everett [1992]. Clearly, each black-colored vertex is adjacent to either another black-colored vertex or a white-colored vertex. An interesting case is formed by the white-colored vertices. Clearly, each such vertex may be joined with a vertex of any color. However, what's important is that for *every* white-colored vertex joined with any vertex of color c, another white-colored vertex will be joined with a vertex of color c as well. This is the essence of being regularly equivalent.

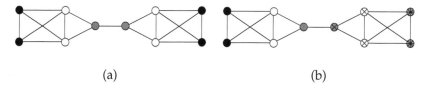

(a) (b)

Figure 9.22: (a) Coloring the vertices of a graph to identify regular equivalence. (b) An alternative coloring that also reflects structural equivalence.

Figure 9.22(b) shows an alternative coloring that also reflects structurally equivalent vertices. In general, if two vertices are structurally equivalent, they will also be regularly equivalent.

Conclusions

We have come a long way, and for some the road has inevitably been rough. Summarizing, there are essentially three major topics that should have been picked up by now:

1. Basic graph theory
2. Metrics for graph analysis
3. Basics of complex network theory

Let's consider each of these briefly.

Graph theory

Chapters 2 to 5 cover the basic material that you would find in most introductory courses on graph theory. We have discussed the foundations of graphs, including their representations and embeddings in the plane, allowing us to speak more accurately about graphs that are the same. Understanding the foundations is important in order to make any steps in understanding real-world networks.

Chapter 3 can be somewhat considered as a collection of randomly collected general topics on graph theory, but which form essential extensions to the foundations discussed in the preceeding chapter. Again, we see that real-world networks can be much easier modeled when edges can be directed or have a weight. When it comes to coloring, notably the vertex coloring proves to come in handy when certain properties of networks need to be shown.

There are many topics that we have not discussed that could easily be categorized as extensions to our foundations. Examples include matchings and independent sets, as well as a discussion of many special graphs. Also important is the topic of network flows which we have completely ignored. There are several more for which it can be argued that they deserve a place in any introductory book on graph theory. However, many of these topics are less relevant in light of understanding real-world networks. Instead, they often turn out to be particularly useful in the context of optimization problems, which lies at the heart of a field of mathematics known as operations research.

In this light, one may argue that discussing Euler tours and Hamilton cycles, the topics of Chapter 4, could equally well have been skipped. However, these topics were felt to be so fundamental that skipping them would not have been the right thing to do. Notably the subtle difference between the two concepts is important, and when realizing that the traveling salesman problem alone is whole field of research by itself, and important for information and computer scientists, skipping it is not really an option.

In Chapter 5 we discussed trees, which form a recurring subject in many courses taught for IT students. Notably the issue of routing, i.e., shortest paths, is fundamental to understanding how information may be disseminated through a network.

Graph analysis

When we started to discuss metrics for graphs, we started to deviate from classical texts on graph theory. Although concepts such as eccentricity and the center of a graph can be found in many standard textbooks on graph theory, they are not as much emphasized as they are here. Understanding and being able to say anything sensible about real-world networks requires a thorough understanding of graph-theoretical metrics. It is through these metrics that it becomes possible, for example, to assess the complexity of a network.

It is actually surprising how few metrics are generally used. What should be less surprising is that with discrete structures such as graphs finding metrics that lead to a "soft" classification of networks is much more difficult. For example, finding the center of a graph may not be very useful for a random network, as it may easily consist of only one or very few nodes. More important is that we would be able to identify all the nodes in a center by a more relaxed criterion than having minimal eccentricity. This is especially true for large networks, but it may also hold for relatively small networks.

We bumped into this phenomenon a few times, notably in the case of social networks when discussing cohesive subgroups and later structural equivalence. What is needed are metrics that can adequately capture the fuzziness of what we tend to call in our daily lives "cliques," the "most important" people or organizations, and so on. This book has only briefly touched upon some of the attempts at grasping such metrics.

Complex networks

Finally, we have completely deviated from standard introductory textbooks on graph theory with the material covered by Chapter 7 through 9, with the exception of some material about social networks.

Complex networks in many senses capture what we can observe in real life and for that reason alone they are important to study. Graph theory forms the foundations for understanding complex networks and the first part of the book should be sufficient to make a next step.

Chapter 7 provides the fundamentals for going into the real-world networks discussed later in the book. The three types of random networks—Erdös-Rényi graphs, Watts-Strogatz graphs, and scale-free networks—form the basis for classifying real-world networks. Notably the combination of small-world properties and high clustering as witnessed in the case of scale-free networks is important.

We actually discussed only very few real-world complex networks with the Internet and Web from Chapter 8 being the most illustrative, along with the structured and unstructured peer-to-peer networks. There are many examples of complex networks to be found in fields such as transportation, neurology, biology, financial markets, language, etc. With the examples given in Chapter 8 it should not be too difficult to take further steps in understanding such networks. Again, it is important to consider how exactly complex networks are measured. We encountered that in many cases the networks can be so large that we need to resort to sampling techniques, which immediately brings up the problem of data validity. In other words, is our sample good enough to be representative for the entire network? We saw that in the case of the Web, answering this question may be far from trivial.

Social networks, discussed in Chapter 9, in many ways brought us back to more traditional graph theory. One could consider many social network tools to form an extension to graph theory as discussed in Chapter 2, but targeted to a specific field of study. What we have not discussed is the link between traditional social networks and social online communities, an emerging subdiscipline in the field of network science. As may be expected, social online communities exhibit many of the properties common to complex networks, yet at the same time become interesting when we attempt to discover social structures. Again, with the material covered in the second half of the book, the reader should be able to easily follow the literature on complex social networks.

Next steps and suggested textbooks

After having ploughed through this book, a wide range of topics lie open for further exploration. In the first place, for those who have become more interested in math and graph theory, there are many excellent textbooks that can be picked up from here. A good starting point is formed by West [2001], although many will still appreciate the somewhat outdated, yet excellent work described in Bondy and Murty [1976]. Another good and certainly gentle introduction is provided by Aldous and Wilson [2000], who put much less emphasis on formal notations than we have done. When it comes to understanding proofs and mathematical notations, Velleman [2006] is highly recommended.

There are few topics in graph theory to which entire books have been devoted. Some of the ones that will now make a lot of sense and will certainly be appreciated are the following two. First, Wilson [2004] provides a very nice and interesting historical read on the four-color problem. The importance of the traveling salesman problem should have become clear by now. An excellent description of what it takes to put it to practice is described in Applegate et al. [2007], although it does require going through some more serious math.

There are not many books that concentrate entirely on network analysis. Brandes and Erlebach [2005] contains a collection of articles that describe different aspects of network analysis, including a chapter on various metrics. It may be useful as background material and at the same time a concise reference to various graph-theoretical foundations.

When it comes to complex networks, an excellent starting point is formed by Barabási [2002], an exceptionally well-written book that will mostly likely trigger further interest into the topic. Equally recommended is Watts [2003] which also concentrates on complex networks. Complexity in general is discussed in Mitchell [2009], a very accessible read into the fascinating field of what is called complexity research. When it comes to getting an overview of all the important publications that gave form to the research into complex networks, Newman et al. [2006] is the place to go. This edited book is a collection of original papers on random graphs, scale-free networks, the structure of the Web and so on. Going through some serious math is sometimes needed, but rewarding.

Finally, for those who have picked up an interest in social network analysis, a good point to start is Knoke and Yang [2008]. Scott [2000] provides an excellent overview on different topics, treating them in independent chapters. The definitive guide to social network analysis, however, remains Wasserman and Faust [1994]. An extensive, yet reasonably accessible piece of work. Finally, going from structure to content, Christakis and Fowler [2009] concentrate less on the structure of social networks and instead attempt to discover and explain the meaning behind links in social networks and what information can be derived from those networks.

Mathematical Notations

	Basic set notations
$\mathbb{N}$	The set of natural numbers.
$\mathbb{R}$	The set of real numbers.
$|S|$	The size of a (finite) set S.
$\min S$	The smallest value found in set S.
$\max S$	The largest value found in set S.
$\forall$	The universal quantifier, used in statements such as "for all ...".
$\exists$	The existential quantifier, used in statements such as "there exists ...".
$x \in S$	Element x is a member of set S.
$V \setminus W$	The set V excluding elements that are also member of W.
$V \subseteq W$	Denotes that the set V is a subset of W, and possibly equal to W.
$V \subset W$	Denotes that V is a proper subset of W, i.e., $V \subseteq W$ and $V \neq W$.
$V \cap W$	The intersection of the two sets V and W.
$\bigcap_{i=1}^{n} V_i$	The intersection of n sets: $V_1 \cap V_2 \cap \cdots \cap V_n$
$V \cup W$	The union of the two sets V and W.
$\bigcup_{i=1}^{n} V_i$	The union of n sets: $V_1 \cup V_2 \cup \cdots \cup V_n$
	General mathematical notations
$\lceil x \rceil$	The smallest natural number greater or equal to x.
$\lfloor x \rfloor$	The largest natural number smaller or equal to x.
$n!$	To be pronounced as n factorial: $n! \stackrel{\text{def}}{=} n \cdot (n-1) \cdot (n-2) \cdots 1$.
$n \gg k$	The fact that n is much larger than k.

Σ	Summation, such as $\sum_{i=1}^{n} x_i$, meaning $x_1 + x_2 + \cdots + x_n$.
Π	Multiplication, such as $\Pi_{i=1}^{n} x_i$, meaning $x_1 \times x_2 \times \cdots \times x_n$.
$[a_1, a_2, \ldots a_n]$	The (ordered) sequence of elements $a_1, a_2, \ldots, a_n$.
$x \leftarrow S$	x takes the value resulting from the expression S, pronounced as "x becomes S".
$f(x) \sim \mathcal{O}(g(x))$	$f(x)$ is bounded by $g(x)$: $\exists M \, \forall x > x_0 : |f(x)| < M \cdot |g(x)|$
$f(x) \sim \Omega(g(x)$	$f(x)$ is bounded from below by $g(x)$: $\exists M \, \forall x > x_0 : |f(x)| > M \cdot |g(x)|$. This also means that $g(x) \sim \mathcal{O}(f(x))$.
$f(x) \sim \Theta(g(x))$	$f(x)$ follows the same form as $g(x)$: $\exists M, M' \, \forall x > x_0 : M'|g(x)| < |f(x)| < M|g(x)|$.

General graph-theory notations	
$G = (V, E)$	The undirected graph G with vertex set V and edge set E.
$\langle u, v \rangle$	The fact that vertex u and v are joined by an edge, that is, they are adjacent.
$\neg \langle u, v \rangle$	The fact that vertex u and v are *not* adjacent.
$D = (V, A)$	The directed graph D with vertex set V and arc set A.
$\langle \overrightarrow{u, v} \rangle$	The fact that vertex u and v are joined by an arc *from u to v*.
$G[V^*]$	The graph induced by the set of vertices $V^* \subseteq V(G)$.
$G[E^*]$	The graph induced by the set of edges $E^* \subseteq E(G)$.
$H \subseteq G$	H is a subgraph of G.
$G - v$	The graph induced by $V(G) \setminus \{v\}$.
$G - e$	The graph induced by $E(G) \setminus \{e\}$.
K_n	The complete graph on $n > 0$ vertices.
$K_{m,n}$	The complete bipartite graph with with two vertex sets of size m and n, respectively.
$\overline{G}$	The complement of graph G, i.e., the graph obtained from G by removing its edges and joining vertices that were nonadjacent in G.
$H_{k,n}$	A k-connected graph with n vertices and a minimal number of edges: a Harary graph.
$N(v)$	The set of neighbors of vertex v.
$N_{in}(v)$	The set of in-neighbors of vertex v.
$N_{out}(v)$	The set of out-neighbors of vertex v.

$\delta(v)$	The degree of vertex v, i.e., the number of incident edges.
$\delta_{in}(v)$	The indegree of vertex v, i.e., the number of incoming arcs at v.
$\delta_{out}(v)$	The outdegree of vertex v, i.e., the number of outgoing arcs from v.
$\Delta(G)$	The maximal degree of any vertex in graph G: $\max\{\delta(v)|v \in V(G)\}$.

Metrics on graphs

$d(u,v)$	The geodesic distance between vertex u and v. This is either a minimal-length (u,v)-path or a minimal-weight (u,v)-path..
$\epsilon(u)$	The eccentricity of vertex u: the maximum distance of u to any other vertex.
$\tau(G)$	The network transitivity of graph G: the ratio between the number of triangles and triples in G.
$c_C(u)$	The closeness of vertex u (in a graph G), measured as the reciproke of the total distance u has to the other vertices of G.
$c_B(u)$	The betweenness centrality of vertex u: the ratio of shortest paths between two vertices that go through u.
$c_E(u)$	The vertex centrality of u: the reciproke of its eccentricity.
$diam(G)$	The diameter of graph G: the length of the longest shortest path between any two vertices, i.e., the maximal eccentricity among the vertices of G.
$rad(G)$	The radius of graph G: the minimal eccentricity among its vertices.
$C(G)$	The center of graph G: the set of vertices for which the eccentricity is the same as the radius of G.
$cc(v)$	The clustering coefficient of vertex v.
$CC(G)$	The average clustering coefficient measured over all vertices of graph G.
$\omega(G)$	The number of components of graph G.
$\kappa(G)$	The size of a minimal vertex cut of graph G.
$\lambda(G)$	The size of a minimal edge cut of graph G.
$\chi'(G)$	The edge chromatic number of G: the minimal k for which graph G is k-edge colorable.
$\chi(G)$	The chromatic number of G: the minimal k for which graph G is k-vertex colorable.

	Probabilities
$\mathbb{P}[\delta = k]$	The probability that the degree (of an arbitrarily chosen vertex) is equal to k.
$P[k]$	An abbreviation for $\mathbb{P}[\delta = k]$.
$\mathbb{E}[X]$	The expected value of the random variable X (often corresponding to the *mean*).

	Special classes of graphs
$ER(n, p)$	The collection of Erdös-Rényi random graphs with n vertices and probability p that two distinct vertices are joined.
$WS(n, k, p)$	The collection of Watts-Strogatz random graphs with n vertices, initial vertex degree k and rewiring probability p.
$BA(n, n_0, m)$	The collection of Barabási-Albert random graphs with n vertices, n_0 initial vertices and a growth of m edges at each step.

INDEX

k-regular graph, 23

access network, 191
acyclic graph, 51
address, 188
    MAC, 189
address, host identifier, 190
address,IP, 189
address,network identifier, 190
adjacency matrix, 31, 138
    symmetric, 31
adjacent vertices, 19
ADSL connection, 103
algorithm
    breadth first, 62
arc, 57
    head, 58
    tail, 58
AS, *see* autonomous system
AS number, 192
AS topology, 192
assortative mixing, 137
automorphic equivalence, *see* equivalence
automorphism, 258
autonomous system, 192
average path length, 141

BA graph, *see* random graph

Barabási-Albert graph, *see* random graph
Bellman-Ford algorithm, 124
betweenness centrality, 152, 227, 233, 234
BGP, *see* Border Gateway Protocol
big O notation, 128
binomial distribution, 159
bipartite graph, 46
    complete, 53
block modeling, 229
border gateway, *see* gateway, border
Border Gateway Protocol, 193
bowtie, *see* Web graph

center of a graph, 151
characteristic path length, 141
Chinese postman problem, 87
Chord, 197
    finger table, 199
    successor, 198
chromatic number, 72
circular embedding, 45
client, 195
client-server architecture, 195
clique
    adjacent k-cliques, 249
    community, 249
    directed k-clique, 251

k-clan, 247
k-clique, 246
k-club, 248
k-distance-clique, 247
maximal, 246
clique percolation, 249
closed walk, 37
closed walk, 83
closeness, 151, 233, 234
clustering
   global view, 147
   local view, 144
clustering coefficient
   of a vertex, 164
   of a directed graph, 145
   of a graph, 144
   of a vertex, 144, 145
   of a vertex in a weighted graph, 145
cohesive subgroup, 246
communication
   heliographic, 4
   telegraphic, 4
communication protocol, 5
complete bipartite graph, 53
complete graph, 19
complex network, 3
component, 38
computationally efficient, 130
computationally inefficient, 130
connected world, 4
connected graph, 37
connected vertices, 37
connectivity
   k-connected, 39
   k-edge-connected, 39
   optimally connected, 39
connector problem, 107
correlation coefficient, 137
count-to-infinity problem, 127
cubic graph, 23, 29

curve fitting, 136
cut edge, 38
cut vertex, 38, 152
cycle, 37
   directed, 61
cycle time, *see* epidemic protocol

DAG, *see* directed acyclic graph
decentralized algorithm, 126
degree correlation, 138
degree correlation, 137
degree distribution
   power law, 173
degree prestige, 235
degree sequence, 23
   ordered, 23
DHCP, *see* Dynamic Host Configuration Protocol
DHCP server, 189
diameter, 141
digraph, 57
   strongly connected, 61
   weakly connected, 61
Dijkstra's algorithm, 120
Dirac's theorem, 95
direct proof, 73
directed cycle, 61
directed walk, 61
directed acyclic graph, 61
directed graph, 57
   acyclic, 61
   arc, 57
   orientation, 58
   strict, 58
directed k-clique, *see* clique
directed path, 61
directed trail, 61
DISCONNECTED, *see* Web graph
disconnected graph, 38
distance
   between vertices, 46, 66
   Euclidean, 256

INDEX

geodesic, 66, 140
DNS, *see* Domain Name System
Domain Name System, 213
domain name, 212
Dynamic Host Configuration Protocol, 189

eccentricity, 140, 151, 233
edge, 10, 18
    duplicating, 89
    end point, 19
    incident, 19
    loop, 19
    multiple, 19, 69
    weight, 65
edge list, 33
edge chromatic number, 71
edge coloring, 71
    minimal, 70
edge cut, 38
edge-independent paths, 40
eigenvalue, 239
eigenvector, 238, 239
end point, *see* edge,end point
epidemic dissemination, 143
epidemic protocol, *see* peer-to-peer
    cycle time, 206, 208
    round, 208, 209
epidemic-based network, 204
equivalence
    automorphic, 259
    regular, 260
equivalence, structural, 255
ER random graph, *see* random graph
Euclidean distance, *see* distance,Euclidean
Euler constant, 162, 180
Euler tour, 83
Euler trail, 84
existential quantifier, 20
existential proof, 76, 96
expected value, *see* random variable

finger table, *see* Chord
flow of control, 63, 68
forest, 51

gateway, border, 191
geodesic, 66
geodesic distance, *see* distance
giant component, 165
gossiping, *see* epidemic-based networks
gossiping models, 143
graph, 10
    k-regular, 23
    acyclic, 51
    automorphism, 258
    center, 151
    complement, 19
    component, 38
    connected, 37
    definition, 18
    directed, 57
    disconnected, 38
    edge, 18
    empty, 19
    Hamiltonian, 92
    induced, 29
    isomorphism, 33, 258
    join vertices, 18
    line, 30
    orientation, 58
    planar, 50
    plane, 50
    regular, 23
    simple, 19, 31, 58
    subgraph, 29
    tree, *see* tree
    union, 29
    vertex, 18, 57
    weighted, 65
graph embedding
    circular, 45
    ranked, 46

spring, 47
graph closure, 97
graph embedding, 45
graph theory, 13, 18
graphic, 23
grid graph, 127

Hamilton cycle, 81, 92
Hamilton path, 92
Hamiltonian graph, 92
Harary graph, 41
head, *see* arc,head
home network, 190
homophily, 226
host, 187
host identifier, *see* address, host identifier
HTML, *see* HyperText Markup Language
HTTP, *see* HyperText Transfer Protocol
HTTP request, 213
hub, 133
hyperlink, 212, 213
HyperText Markup Language, 214
HyperText Transfer Protocol, 213

iff, 26
IN, *see* Web graph
in-neighbor set, 58
incidence matrix, 31
indegree, 58
independent set, 263
indirect proof, 73
induced graph, 29
infix notation, 110
influence domain, 235
interface, 119
    communication, 119
Internet Protocol, 189
Internet Service Provider, 191
Internet, edge, 191

IP, *see* Internet Protocol
IP address, *see* address,IP
isomorphic graphs, 33, 258
ISP, *see* Internet Service Provider

k-clan, *see* clique,k-clan
k-clique community, *see* clique
k-club, *see* clique,k-club
k-connected graph, 39
k-core, 249
k-distance-clique, *see* clique
k-edge coloring, 71
k-edge connected graph, 39
k-vertex coloring, 71

LAN, *see* local-area network
line graph, *see* graph, line
local-area network, 188
loop, *see* edge,loop
lower bound, 102

MAC address, *see* address,MAC
markup language, 214
matching, 91, 263
    perfect, 92
MBone, 107
mean (of a random variable), *see* random variable
median, 141
Menger, Karl, 40
message routing, 119
multiple arc, 69
multiple edge, *see* edge,multiple

neighbor set, 20
network
    transportation, 107
network transitivity, 149
network science, 10
network density, 147, 163
    sparse, 168
network flow, 263

network identifier, *see* address, network identifier
network science, 11
network transitivity, 147
network, access, 191
network, home, 190
network, tier 1, 192
network, tier 2, 191
network, tier 3, 191
nonconstructive proof, 27

one-mode network, 252
optimally connected graph, 39
orientation, 58
OUT, *see* Web graph
out-neighbor set, 58
outdegree, 58
overlay network, 109, 196

packet, 187
PageRank, 217
partial view, 196
path, 37
    directed, 61
    edge-independent, 40
    length, 126
    vertex-independent, 40
peer, 196
peer-to-peer
    epidemics, 204
peer-to-peer network, 196
    unstructured, 204
peering relationship, 191
perfect matching, 92
Petersen graph, 45
pigeonhole principle, 44
planar graph, 50
planar graph
    exterior region, 50
    face, 50
    interior region, 50
    region, 50

plane graph, 50
Posa's algorithm, 99
power law distribution, *see* degree distribution
preferential attachment, 175
prefix notation, 110
proof technique
    extremality, 84
proof techniques
    existential, 96
proof by contradiction, 44
proof by induction, 51
proof by construction, 27, 96
proof techniques
    by construction, 27, 96
    by contradiction, 44
    by induction, 51
    direct, 73
    existential, 76
    extremality, 96
    indirect, 73
proximity prestige, 235
pseudo-code, 63
    control flow, 63

radius, 140
random variable, 159
random graph, 45
    Barabási-Albert, 175
    ER random graph, 158
    Erdös-Rényi graph, 158
    Watts-Strogatz, 168
random network
    see random graph, 158
random variable, 159
    discrete, 159
    expected value, 160
    mean, 160
ranked embedding, 46
ranked prestige, 236
reachability analysis, 62
regular graph, 23

regular equivalence, *see* equivalence
rooted tree, 109, 120
rotational transformation, 99
round, *see* epidemic protocol
router, 188, 189
routing, 119, 187
routing algorithm, 66
routing cost, 124
routing protocol, 119
    distance vector, 126
    link state, 120
routing table, 119

scale-free network, 172
scale-freeness, 139
    normalized, 140
scaling exponent, 173
SCC, *see* Web graph, SCC
server, 195
shortest path, 46, 66
shutter telegraph, 5
sign, 240
    product of, 242
signed graph, 240
    balanced, 243
sink tree, 120
small-world network, 167
social balance, *see* structural balance
social network, 167, 225
sociogram, 10, 227, 231
sociometry, 228
spanning tree, 109
spanning walk, 81
sparse network, *see* network density
spider trap, 216
spring embedding, 47
standard deviation, 138
strict, *see* directed graph,strict
strongly connected digraph, 61

structural equivalence, *see* equivalence
structural balance, 228, 240
subgraph, 29
super small world, 181
surface Web, 216
switch, 188

tail, *see* arc,tail
telegraphic communication, 4
TENDRIL, *see* Web graph
topology, 123
tour, 81, 83
trail, 37
    directed, 61
transportation network, 107
traveling salesman problem, 93
tree, 6, 51, 68, 107
    binary, 111
    descendant, 111
    intermediate node, 109
    leaf node, 109
    parent, 111
    rooted, 109, 120
    sink, 120
    spanning, 109
triad, 228, 240
triangle, 146
    at a vertex, 146
    transitive, 150
    weight, 149
triple
    at a vertex, 146
    nonvacuous, 150
    weight, 149
TSP, *see* traveling salesman problem
TUBE, *see* Web graph
two-mode network, 252

underlying graph, 58
Uniform Resource Locator, 213

INDEX

URL, *see* Uniform Resource Locator

vertex, 10, 18, 57
    adjacent, 19
    degree, 21
    degree correlation, 137, 138
    indegree, 58
    outdegree, 58
    type, 137
vertex degree
    distribution, 59
vertex centrality, 151, 234
vertex coloring, 71
vertex cut, 38
vertex degree, 21, 31, 32
    distribution, 22
vertex degree distribution, 59
vertex reachability, 62
vertex strength, 145
vertex-independent paths, 40
virtual network, 42

walk, 37, 81
    closed, 37, 83
    directed, 61
    spanning, 81
Watts-Strogatz random graph, 168
weak link, 169
weak tie, 230
weakly connected digraph, 61
Web subgraph
    bowtie, 217
Web client, 213
Web crawling
    breadth first, 216
    PageRank, 217, 220
    random selection, 217
Web Graph
    SCC (Strongly Connected Component), 217
Web graph, 214

DISCONNECTED, 218
IN, 217
OUT, 218
TENDRIL, 218
TUBE, 218
Web server, 213
Web site, 212
Web subgraph, 217
weight, 65
weighted average, 160
weighted clustering coefficient, 145
weighted graph, 65
World Wide Web, 212
WS random graph, *see* random graph, Watts-Strogatz
WWW, *see* World Wide Web

BIBLIOGRAPHY

Adamic L. The Small World Web. In Abiteboul S. and Vercoustre A.-M., editors, *ECDL*, volume 1696 of *Lecture Notes on Computer Science*, pages 443–452, Berlin, Sept. 1999. Springer-Verlag. Cited on 221, 222

Albitz P. and Liu C. *DNS and BIND*. O'Reilly & Associates, Sebastopol, CA., 4th edition, 2001. Cited on 213

Aldous J. and Wilson R. *Graphs and Applications, An Introductory Approach*. Springer-Verlag, Berlin, 2000. Cited on 265

Appel K. and Haken W. Every Planar Map is Four Colorable. *Bull. Amer. Math. Soc.*, 82:711–712, 1976. Cited on 74

Appel K. and Haken W. The Four Color Proof Suffices. *Mathematical Intelligencer*, 8 (1):10–20, 1986. Cited on 74

Applegate D. L., Bixby R. E., Chvatal V., and Cook W. J. *The Traveling Salesman Problem: A Computational Study*. Princeton University Press, Princeton, NJ, 2007. Cited on 93, 266

Barabási A.-L. *Linked, The New Science of Networks*. Perseus Publishing, Cambridge, MA, 2002. Cited on 157, 266

Barabási A.-L. and Albert R. The Emergence of Scaling in Random Networks. *Science*, 286:797–817, 1999. Cited on 172, 175

Barrat A., Barth'elemy M., Pastor-Satorras R., and Vespignani A. The Architecture of Complex Weighted Networks. *Proceedings of the National Acadamy of Sciences of the United States of America*, 101(11):3747–3752, Mar. 2004. Cited on 145

Barroso L., Deam J., and Holze U. Web Search for a Planet: The Google Cluster Architecture. *IEEE Micro*, 23(2):21–28, Mar. 2003. Cited on 215

Becchetti L., Castillo C., Donato D., and Fazzone A. A Comparison of Sampling Techniques for Web Graph Characterization. In *Workshop on Link Analysis: Dynamics and Static of Large Networks (LinkKDD2006)*, New York, NY, Aug. 2006. ACM, ACM Press. Cited on 216

Boldi P., Santini M., and Vigna S. PageRank as a Function of the Damping Factor. In *14th International World Wide Web Conference*, pages 557–566, New York, NY, May 2005. ACM, ACM Press. Cited on 221

Bondy J. and Murty U. *Graph Theory with Applications*. Macmillan, London, 1976. Cited on 42, 71, 85, 118, 265

Bondy J. and Murty U. *Graph Theory*. Springer-Verlag, Berlin, 2008. Cited on 41

Borgatti S. and Everett M. The Notion of Position in Social Network Analysis. *Sociological Methodology*, 22:1–35, 1992. Cited on 260

Brandes U. and Erlebach T., editors. *Network Analysis: Methodological Foundations*, volume 3418 of *Lecture Notes on Computer Science*. Springer-Verlag, Berlin, 2005. Cited on 133, 266

Brin S. and Page L. The Anatomy of a Large-scale Hypertextual Web Search Engine. *Computer Networks*, 30(1-7):107–117, 1998. Cited on 217

Brinkmeier M. and Shank T. Network Statistics. In Brandes U. and Erlebach T., editors, *Network Analysis*, volume 3418 of *Lecture Notes on Computer Science*, pages 293–317. Springer-Verlag, Berlin, 2005. Cited on 140

Broder A., Kumar R., Maghoul F., Raghavan P., Rajagopalan S., Stata R., Tomkins A., and Wiener J. Graph Structure in the Web. *Computer Networks*, 33(1-6):309–320, 2000. Cited on 217, 218, 219, 222

Buchanan M. *Nexus: Small Worlds and the Groundbreaking Science of Networks*. Norton, New York, NY, 2002. Cited on 157

Cartwright D. and Harary F. Structural Balance: A Generalization of Heider's Theory. *Psychological Review*, 63(5):277–293, 1956. Cited on 228

Chartrand G. *Introductory Graph Theory*. Dover Publications, New York, NY, 1977. Cited on 76

Chi Y.-J., Oliveira R., and Zhang L. Cyclops: The AS-level Connectivity Observatory. *ACM Computer Communications Review*, Oct. 2008. Cited on 194, 195

Cho J., Garcia-Molina H., Haveliwala T., Lam W., Paepcke A., Raghavan S., and Wesley G. Stanford WebBase Components and Applications. *ACM Transactions on Internet Technology*, 6(2):153–186, 2006. Cited on 219

Christakis N. and Fowler J. *Connected: The Surprising Power of Our Social Networks and How They Shape Our Lives*. Little, Brown and Company, New York, NY, 2009. Cited on 266

Cook D. and Holder L., editors. *Mining Graph Data*. John Wiley, New York, 2007. Cited on 133

Cothey V. Web-crawling Reliability. *Journal of the American Society for Information Science and Technology*, 55(14):1228–1238, 2004. Cited on 219

Csermely P. *Weak Links: Stabilizers of Complex Systems from Proteins to Social Networks*. Springer-Verlag, Berlin, 2006. Cited on 169

d'Angelo J. and West D. *Mathematical Thinking, Problem-Solving and Proofs*. Prentice Hall, Englewood Cliffs, N.J., 2nd edition, 2000. Cited on 51

Nooy W.de . Description of the data. Technical note, Nov. 2006. http://home.medewerker.uva.nl/w.denooy/. Cited on 254

Nooy W.de , Mrvar A., and Batagelj V. *Exploratory Social Network Analysis with Pajek*. Cambridge University Press, Cambridge, UK, 2005. Cited on 225, 230

Dekker W. and Raaij B.van . *De Elite*. Meulenhoff, Amsterdam, The Netherlands, 2006. In Dutch. Cited on 253

Demers A., Greene D., Hauser C., Irish W., Larson J., Shenker S., Sturgis H., Swinehart D., and Terry D. Epidemic Algorithms for Replicated Database Maintenance. In *6th Symposium on Principles of Distributed Computing*, pages 1–12. ACM, Aug.

1987. Cited on 206

Dharwadker A. A New Algorithm for finding Hamiltonian Circuits. http://www.geocities.com/dharwadker/hamilton, 2004. Cited on 96

Diestel R. *Graph Theory.* Springer-Verlag, Berlin, 3rd edition, 2005. Cited on 41

Dodds P., Muhammed R., and Watts D. J. An Experimental Study of Search in Global Social Networks. *Science*, 301:827–829, Aug. 2003. Cited on 10

Donato D., Laura L., Leonardi S., and Millozzi S. The Web as a Graph: How far we are. *ACM Transactions on Internet Technology*, 7(1), Jan. 2007. Cited on 220, 221

Dorogovtsev S., Mendes J., and Samukhin A. Structure of Growing Networks with Preferential Linking. *Physical Review Letters*, 85:4633–4636, 2000. Cited on 175

Dorogovtsev S., Mendes J., and Dorogovtsev S. *Evolution of Networks: From Biological Nets to the Internet and WWW.* Oxford University Press, New York, NY, 2003. Cited on 174, 178, 182

Dunne J., Williams R., and Martinez N. Food-Web Structure and Network Theory: The Role of Connectance and Size. *Proceedings of the National Acadamy of Sciences of the United States of America*, 99(20):12917–12922, Oct. 2002. Cited on 158

Eades P. A Heuristic for Graph Drawing. *Congressus Numerantium*, 42:149–160, 1984. Cited on 47, 48

Edmonds J. and Johnson E. Matching, Euler Tours, and the Chinese Postman. *Mathematical Programming*, 5(1):88–124, Dec. 1973. Cited on 90

Erdös P. and Rényi A. On Random Graphs. *Publicationes Mathematicae*, 6:290–297, 1959. Cited on 158

Eriksson H. MBone: The Muliticast Backbone. *Communications of the ACM*, 37(8):54–60, Aug. 1994. Cited on 107

Eugster P., Guerraoui R., Kermarrec A.-M., and Massoulié L. Epidemic Information Dissemination in Distributed Systems. *Computer*, 37(5):60–67, May 2004. Cited on 143

Fronczak A., Fronczak P., and Holyst J. Mean-field Theory for Clustering Coefficients in Barabási-Albert Networks. *Physical Review E*, 68(4):046126, Oct. 2003. Cited on 178

Fronczak A., Fronczak P., and Holyst J. Average Path Length in Random Networks. *Physical Review E*, 70(5):056110, Nov. 2004. Cited on 162, 179

Garey M. and Johnson D. *Computers and Intractibility: A Guide to the Theory of NP-Completeness.* Freeman, New York, 1979. Cited on 130

Gibbons A. *Algorithmic Graph Theory.* Cambridge University Press, Cambridge, UK, 1985. Cited on 90, 92

Goodrich M. and Tamassia R. *Algorithm Design: Foundations, Analysis and Internet Examples.* John Wiley, New York, 2002. Cited on 112, 121, 129

Graham R. and Hell P. On the History of the Minimum Spanning Tree Problem. *Annals of the History of Computing*, 7(1):43–57, Jan. 1985. Cited on 116

Granovetter M. The Strength of Weak Ties. *American Journal of Sociology*, 78(6):1360–1380, May 1973. Cited on 230

Grötschel M. and Padberg M. Ulysses 2000: In Search of Optimal Solutions to Hard Combinatorial Problems. Technical Report ZIB-SC-93-34, ZIB, Berlin, Nov. 1993. Cited on 93, 94

Gulli A. and Signorini A. The Indexable Web is More than 11.5 Billion Pages. In *14th International World Wide Web Conference*. ACM, May 2005. Cited on 8

Haddadi H., Fay D., Jamakovic A., Maennel O., Moore A. W., Mortier R., Rio M., and Uhlig S. Beyond Node Degree: Evaluating AS Topology Models. Technical Report UCAM-CL-TR-725, University of Cambridge, Computer Laboratory, Cambridge, UK, July 2008. Cited on 194

Hage P. and Harary F. *Structural Models in Anthropology*. Cambridge University Press, Cambridge, UK, 1983. Cited on 147

Hall J., Hartline J. D., Karlin A. R., Saia J., and Wilkes J. On Algorithms for Efficient Data Migration. In *12th Symposium on Discrete Algorithms*, pages 620–629, New York, NY, Jan. 2001. ACM-SIAM, ACM Press. Cited on 69

Hanneman R. and Riddle M. Introduction to Social Network Methods. Lecture Notes, University of California at Los Angeles, CA, 2005. Cited on 259

Harary F. On the Notion of Balance of a Signed Graph. *Michigan Mathematical Journal*, 2(2):143–146, 1953. Cited on 243, 244

Holme P. and Kim B. Growing Scale-Free Networks with Tunable Clustering. *Pysical Review E*, 65(2):026107, Jan. 2002. Cited on 182, 183

Holzmann G. and Pehrson B. *The Early History of Data Networks*. IEEE Computer Society Press, Los Alamitos, CA., 1995. Cited on 4

Huston G. Exploring Autonomous System Numbers. *The Internet Protocol Journal*, 9 (1):2–23, Mar. 2006. Cited on 193

Jackson M. *Social and Economic Networks*. Princeton University Press, Princeton, NJ, 2008. Cited on 227

Jelasity M., Voulgaris S., Guerraoui R., Kermarrec A.-M., and Steen M.van . Gossip-based Peer Sampling. *ACM Transactions on Computer Systems*, 25(3), Aug. 2007. Cited on 208

Jelasity M., Kowalczyk W., and Steen M.van . Newscast Computing. In *Advanced Computational Technologies*. Romanian Academic Press, 2010. Cited on 209

Jenkins K. and Demers A. Logarithmic Harary Graphs. In *21st International Conference on Distributed Computing Systems Workshops*, Los Alamitos, CA., Apr. 2001. IEEE, IEEE Computer Society Press. Cited on 42

Judd C., McClelland G., and Ryan C. *Data Analysis, A Model Comparison Approach*. Routledge, Hove, UK, 2nd edition, 2009. Cited on 136, 138

Kleinberg J. The Convergence of Social and Technological Networks. *Communications of the ACM*, 51(11):66–72, Nov. 2008. Cited on 230

Knoke D. and Yang S. *Social Network Analysis*. Number 07-154 in Quantative Applications in the Social Sciences. SAGE Publications, Thousand Oaks, CA, 2nd edition, 2008. Cited on 252, 266

Kotschutzki D., Lehmann K., Peeters L., Richter S., Tenfelde-Podehl D., and Zlotowski O. Centrality Indices. In Brandes U. and Erlebach T., editors, *Network Analysis*, volume 3418 of *Lecture Notes on Computer Science*, pages 16–61. Springer-Verlag, Berlin, 2005. Cited on 151

Kruskal J. On the Shortest Spanning Subtree of a Graph and the Traveling Salesman Problem. *Proc. American Mathematical Society*, 7(1):48–50, Feb. 1956. Cited on 116

Kuan M.-K. Graphic Programming Using Odd or Even Points. *Chinese Mathematics*,

1:273–277, 1962. Cited on 87

Levien R., editor. *Signposts in Cyberspace: The Domain Name System and Internet Navigation*. National Academic Research Council, Washington, DC, 2005. Cited on 213

Lewis T. G. *Network Science: Theory and Practice*. John Wiley, New York, 2009. Cited on 157

Li L., Alderson D., Doyle J., and Willinger W. Towards a Theory of Scale-Free Graphs: Definitions, Properties, and Implications. *Internet Mathematics*, 2(4):431–523, 2005. Cited on 139, 140

Licklider J. and Taylor R. The Computer as a Communication Device. *Science and Technology*, Apr. 1968. Cited on 9

Liu B. *Web Data Mining*. Springer-Verlag, Berlin, 2007. Cited on 215

Lorrain F. and White H. Structural Equivalence of Individuals in Social Networks. *Journal of Mathematical Sociology*, 1:49–80, 1971. Cited on 255

Lua E., Crowcroft J., Pias M., Sharma R., and Lim S. A Survey and Comparison of Peer-to-Peer Overlay Network Schemes. *IEEE Communications Surveys & Tutorials*, 7(2):22–73, Apr. 2005. Cited on 197

Luks E. Isomorphism of Graphs of Bounded Valence can be Tested in Polynomial Time. *Journal of Computer and System Sciences*, 25(1):42–65, Aug. 1982. Cited on 37

Macedonia M. and Brutzman D. MBone Provides Audio and Video Across the Internet. *Computer*, 27(4):30–36, Apr. 1994. Cited on 107

Malkin G. and Steenstrup M. Distance-Vector Routing. In Steenstrup M., editor, *Routing in Communications Networks*, pages 83–98. Prentice Hall, Englewood Cliffs, N.J., 1995. Cited on 126

Mandel J. *The Statistical Analysis of Experimental Data*. Dover Publications, New York, NY, 1984. Cited on 138

McKay B. Practical Graph Isomorphism. *Congressus Numerantium*, 30:45–87, 1980. Cited on 37

McQuillan J. Graph Theory Applied to Optimal Connectivity in Computer Networks. *ACM Computer Communications Review*, 7(2):13–41, Apr. 1977. Cited on 42

Michael J. Labor Dispute Reconciliation in a Forest Products Manufacturing Facility. *Forest Products Journal*, 47(11):41–45, Nov. 1997. Cited on 225

Mitchell M. *Complexity, A Guided Tour*. Oxford University Press, Oxford, UK, 2009. Cited on 266

Mokken J. Cliques, Clubs, and Clans. *Quality and Quantity*, 13(2):161–173, Apr. 1979. Cited on 246, 247

Moy J. Link-State Routing. In Steenstrup M., editor, *Routing in Communications Networks*, pages 135–157. Prentice Hall, Englewood Cliffs, N.J., 1995. Cited on 120

Newman M. Assortative Mixing in Networks. *Physical Review Letters*, 89:208701, 2002. Cited on 136

Newman M. The Structure and Function of Complex Networks. *SIAM Review*, 45: 167–256, 2003a. Cited on 146

Newman M. Mixing Patterns in Networks. *Phys. Rev. E*, 67(2):026126, Feb. 2003b. Cited on 137

Newman M., Barabasi A.-L., and Watts D., editors. *The Structure and Dynamics of*

Networks. Princeton University Press, Princeton, NJ, 2006. Cited on 266

Oliveira R., Zhang B., and Zhang L. In Search of the Elusive Ground Truth: The Internet's AS-level Connectivity Structure. In *International Conference on Measurements and Modeling of Computer Systems*, pages 217–228, New York, NY, June 2008. ACM, ACM Press. Cited on 195

Opsahl T. and Panzarasa P. Clustering in Weighted Networks. *Social Networks*, 31: 155–163, 2009. Cited on 149

Padgett J. and Ansell C. Robust Action and the Rise of the Medici, 1400–1434. *American Journal of Sociology*, 98(6):1259–1319, May 1993. Cited on 226, 227

Palla G., Derényi I., Farkas I., and Vicsek T. Uncovering the Overlapping Community Structure of Complex Networks in Nature and Society. *Nature*, 435:814–818, June 2005. Cited on 249

Palla G., Farkas I., Pollner P., Derényi I., and Vicsek T. Directed Network Modules. *New Journal of Physics*, 9:186, June 2007. Cited on 251, 252

Pandurangan G., Prabhakar R., and Upfal E. Using PageRank to Characterize Web Structure. *Internet Mathematics*, 3(1):1–20, 2006. Cited on 221

Porter M., Onnela J.-P., and Mucha P. Communities in Networks. *Notices of the American Mathematical Society*, 56(9):1082–1097, 2009. Cited on 229, 249

Posa L. Hamiltonian Circuits in Random Graphs. *Discrete Mathematics*, 14(4):359–364, 1976. Cited on 99

Raz D. and Cohen R. The Internet Dark Matter: On The Missing Links in the AS Connectivity Map. In *25th INFOCOM Conference*, pages 1–12, Los Alamitos, CA., Apr. 2006. IEEE, IEEE Computer Society Press. Cited on 195

Salus P. and Quarterman J. Disruptions and Emergencies on the Internet. Matrix NetSystems, Sept. 2002. Talk presented at TPRC 2002. Cited on 3

Scott J. *Social Network Analysis, A Handbook*. SAGE Publications, London, UK, 2nd edition, 2000. Cited on 247, 266

Serrano M., Maguitman A., Boguna M., Fortunato S., and Vespignani A. Decoding the Structure of the WWW: A Comparative Analysis of Web Crawls. *ACM Transactions on the Web*, 1(2), 2007. Cited on 217, 219

Sherman L. Sociometry in the Classroom: How to do it. http://www.users.muohio.edu/shermalw/, 2000. Cited on 231

Stoica I., Morris R., Liben-Nowell D., Karger D. R., Kaashoek M. F., Dabek F., and Balakrishnan H. Chord: A Scalable Peer-to-peer Lookup Protocol for Internet Applications. *IEEE/ACM Transactions on Networking*, 11(1):17–32, Feb. 2003. Cited on 197, 202, 204

Tanenbaum A. and Steen M.van . *Distributed Systems, Principles and Paradigms*. Prentice Hall, Upper Saddle River, N.J., 2nd edition, 2007. Translations: German, Portugese, Italian. Cited on 196

Thelwall M. *Link Analysis, An Information Science Approach*. Elsevier Academic Press, Amsterdam, The Netherlands, 2004. Cited on 215

Thimbleby H. The Directed Chinese Postman Problem. *Software – Practice and Experience*, 33(11):1081–1096, Sept. 2003. Cited on 89

Urdaneta G., Pierre G., and Steen M.van . Wikipedia Workload Analysis for Decentralized Hosting. *Computer Networks*, 53(11):1830–1845, July 2009. Cited on 9

Vandegriend B. Finding Hamiltonian Cycles: Algorithms, Graphs and Performance. Master's thesis, University of Alberta, Department of Computing Science, Edmonton, Canada, 1998. Cited on 99

Vega-Redondo F. *Complex Social Networks*. Cambridge University Press, Cambridge, UK, 2007. Cited on 174, 175

Velleman D. *How To Prove It*. John Wiley, New York, 2nd edition, 2006. Cited on 265

Voss J. Measuring Wikipedia. In *10th International Conference of the International Society for Scientometrics and Informetrics*, July 2005. Cited on 9

Voulgaris S., Gavidia D., and Steen. M.van . CYCLON: Inexpensive Membership Management for Unstructured P2P Overlays. *Journal of Network and Systems Management*, 13(2):197–217, June 2005. Cited on 211

Wams J. and Steen M.van . Internet Messaging. In Singh M., editor, *Practical Handbook of Internet Computing*, chapter 7, pages 7–1–7–18. CRC Press, Boca Raton, FL, 2004. Cited on 9

Wasserman S. and Faust K. *Social Network Analysis: Methods and Applications*. Cambridge University Press, Cambridge, UK, 1994. Cited on 147, 234, 240, 247, 252, 258, 266

Watts D. *Six Degrees: The Science of a Connected Age*. Norton, New York, NY, 2003. Cited on 10, 266

Watts D. The "New" Science of Networks. *Annual Review of Sociology*, 30:243–270, 2004. Cited on 157

Watts D. *Small Worlds, The Dynamics of Networks between Order and Randomness*. Princeton University Press, Princeton, NJ, 1999. Cited on 158

Watts D. and Strogatz S. Collective Dynamics of Small World Networks. *Nature*, 393: 440–442, June 1998. Cited on 144, 167

West D. *An Introduction to Graph Theory*. Prentice Hall, Englewood Cliffs, N.J., 2nd edition, 2001. Cited on 41, 84, 265

Williams G. *Linear Algebra with Applications*. Jones and Bartlett, Sudberry, MA, 4th edition, 2001. Cited on 239

Wilson R. *Four Colors Suffice, How the Map Problem Was Solved*. Princeton University Press, Princeton, NJ, 2004. Cited on 266

Xu W. and Liu Z. How Community Structure Influences Epidemic Spread in Social Networks. *Physica A: Statistical Mechanics and its Applications*, 387(2-3):623–630, Jan. 2008. Cited on 143

Printed in Great Britain
by Amazon.co.uk, Ltd.,
Marston Gate.